United States
Higher Education
and National Security

Edited by Edward L. Mienie

UNG
UNIVERSITY of
NORTH GEORGIA
THE MILITARY COLLEGE OF GEORGIA

INSTITUTE FOR LEADERSHIP
AND STRATEGIC STUDIES

ILSS Symposium Proceedings Series

Copyright © 2023 by University of North Georgia Press

All rights reserved. No part of this book may be reproduced in whole or in part without written permission from the publisher, except by reviewers who may quote brief excerpts in connections with a review in newspaper, magazine, or electronic publications; nor may any part of this book be reproduced, stored in a retrieval system, or transmitted in any form or by any means electronic, mechanical, photocopying, recording, or other, without the written permission from the publisher.

Published by:
University of North Georgia Press
Dahlonega, Georgia

Printing Support by:
Lightning Source Inc.
La Vergne, Tennessee

Cover and book design by Corey Parson.

ISBN: 978-1-959203-03-2

Printed in the United States of America

For more information, please visit: http://ung.edu/university-press
Or e-mail: ungpress@ung.edu

If you need this document in an alternate format for accessibility purposes (e.g. Braille, large print, audio, etc.), please contact Corey Parson at corey.parson@ung.edu or 706-864-1556.

TABLE OF CONTENTS

Preface vii
    Dr. Edward Mienie

Symposium Transcript: General Bob Brown 1
    General Bob Brown

How Higher Education Fills the Security Gap in the Post-Cold War Era 15
    Panelists:
    Dr. Craig Greathouse
    Dr. Cristian Harris
    Dr. Edward Mienie
    Dr. Dlynn Armstrong-Williams
    Dr. Bryson Payne
    Moderator:
    Dr. Daniel S. Papp

Public Diplomacy and International Education 42
    Dr. Dlynn A. Williams and Dr. Cristian Harris

RASCLS vs Ransomware: A Counterintelligence Approach to Cybersecurity Education 68
    Dr. Bryson R. Payne, Dr. Edward L. Mienie, and Obulu J. Anetor

Symposium Transcript: Colonel Wortzel 86
    Colonel Larry M. Wortzel

Rethinking Higher Education Practices to Stimulate Innovation and Global Security 91
    Panelists:
    Professor Jacek Dworzecki
    Associate Professor Izabela Nowicka
    Shannon Vaughn
    Iyonka Strawn-Valcy
    Dr. Magdalena Bogacz
    Dr. Crystal Shelnutt
    Moderator:
    Dr. Michael Lanford

Linking Innovation Back to National Security via Innovation Ecosystems: The Role of Higher Education and Equitable Faculty Socialization 117
    Dr. Magdalena T. Bogacz

Symposium Transcript: Dr. Margaret E. Kosal 142
    Dr. Margaret E. Kosal

Science, Technology, and Strategic Analytics 158
    Panelists:
    Colonel (Retired) Eric Toler
    Lieutenant Colonel Christopher Lowrance
    Dr. J. Sukarno Mertoguno
    Dr. C. Anthony Pfaff
    Moderator:
    Mr. Steven Weldon

Military, Academia, and History: A Vital 21st Century Trinity 184
    Major General Mick Ryan

Leveraging Higher Education to Grow Military Strategists 200
    Panelists:
    Dr. Nicholas Murray
    Colonel Francis Park
    Dr. Robert Davis
    Moderator:
    Dr. Ken Gleiman

Educating Strategists: An Introduction to the Basic Strategic Art Program 225
    Francis J. H. Park

National Security and the Historian's Ethos 243
    Anthony Eames

Encouraging Dual-enrolled Students to Enroll in Corps of Cadets at Senior Military Colleges: Barriers and Opportunities 255
    Dr. Imani Cabell and Dr. Katherine Rose Adams

Higher Education and National Security in the USA .. 265
    Cadet Natali Gvalia

An Overview of Emerging Technologies: A U.S. Coast Guard Cadet Panel 271
    Moderator:
    Dr. Angela G. Jackson-Summers, USCGA
    Cadet Panelists:
    1/c Abby Nitz
    1/c Erin Wood
    2/c Chase Jin
    2/c Michael Dankworth
    2/c Branyelle Carillo
    1/c Nick Epstein

Appendix 287

Preface

Edward Mienie, Ph.D.
University of North Georgia

The University of North Georgia's annual Strategic and Security Studies Symposium, now in its seventh year under the auspices of the Institute for Leadership and Strategic Studies (ILSS), in 2022 examined the role higher education plays within the context of national security interests. The theme for this year's symposium was *United States Higher Education & National Security*. ILSS, the College of Education, and the Strategic and Security Studies Program in collaboration with the US Army War College, the Association of the United States Army, the Army Strategist Association, and the Atlanta Council on International Relations co-sponsored the symposium. The symposium attracted scholars and practitioners from Australia, the Republic of Georgia, Poland, and the United States.

As globalization has brought many positive developments, it has brought some incredible challenges. The world is constantly changing, and we can no longer afford to work in stovepipes. We therefore must deepen our partnership between higher education and the military and non-military elements of national security. The symposium highlighted the educated populace that provides a public and private good for American society, strengthening US military and non-military elements of national security. The US military is a vital component of the national security apparatus and there are numerous means where higher education plays a crucial role in assuring security. A close productive relationship between the US military and higher education is essential to its democratic way of life, and the current relationship is healthy but needs improvement.

The role of higher education in support of US security interests, to enhance cooperation and connectivity between states when governmental efforts are underfunded was discussed. Speakers highlighted the need to prepare strategic leaders for the state security sector as an important mission of the country's higher education system. An argument was made for the importance of curricular innovation at the undergraduate and graduate levels to prepare US students for future careers in diplomacy and security studies. Rethinking higher education practices to stimulate innovation and global security was another important theme that panelists addressed. Speakers posited that the Department of Defense does not have enough in-house capability to either perform the necessary education, training, and certification of its service members to perform critical artificial intelligence (AI) tasks or operate certain AI-driven systems in some cases. They proffered the technological challenge to illustrate where collaborating with industry and academia would be most productive and offered suggestions on overcoming cultural and regulatory barriers to that partnership.

Panelists offered perspectives on the development of military strategists and provided an overview of programs and explanations of the important role that higher education institutions play in developing strategists and the advantages of different approaches. On national security and the historian's ethos, the responsibility of historians who are equally responsible for problems befalling civil-military relations was discussed. Ways in which historians can develop mutually advantageous partnerships with the national security community that benefit American society and revive their profession were suggested.

The nation's colleges and universities could effectively drive innovation and entrepreneurialism focused on national security with their abundance of talented researchers. However, that innovation for national security requires the development and preparation of diverse faculty, as well as the cultivation of meaningful and trusting relationships with the government. One way to do so is through equitable faculty socialization that embraces the distinct advantages inherent in a multicultural society. Panelists discussed barriers to participation of dual enrolled students in the Corps of Cadets with the purpose of combatting those issues and creating innovative solutions to increase participation

Panelists also addressed the rise of new challenges and challengers the US is facing alongside traditional threats to its national security. They argued that, given the uncertainty of where future crises will come from, developing the intellectual capital of the US is a critical investment in its national security. Panelists proposed a novel approach to cybersecurity education that explicitly incorporates counterintelligence principles to help inform users that their information is under attack and that their failure to identify social engineering threats could lead to data breaches. Speakers addressed emerging technologies within the context of security implications and the challenge of mitigating the dual use of these technologies falling into the hands of US adversaries. We need a more robust understanding of the security implications of emerging technologies and a realization that technologies are likely to exacerbate existing social and political divides that may lead to instability. The difference between beneficial and dangerous research is often only one of intent. Panelists made a strong case for how essential it is to engage students in learning about six selected emerging military technologies, namely artificial intelligence, biotechnology, directed energy weapons, hypersonic weapons, lethal autonomous weapons, and quantum technology.

It is our sincere hope that this symposium proceedings will foster further dialog for the improvement of our nation's preparedness to protect our national security interests and well-being, and underscore the important contribution by higher education to that end.

1

SYMPOSIUM TRANSCRIPT
GENERAL BOB BROWN

General Bob Brown

As presented at the 2022 Strategic and Security Studies Symposium
Hosted by the Institute for Leadership and Strategic Studies
University of North Georgia

So, you know a close productive relationship between the US military and higher education is really essential to our democratic way of life and that's a bold statement. But I am absolutely certain that the current relationship is healthy. It's good, through a variety of channels and programs I'll talk about, but we need to improve. It's a complex world, incredibly complex times, and things move at a breakneck speed. Globalization has brought a lot of amazing things; it's also brought some incredible challenges. And so, this relationship between higher education and the military needs to improve. It's essential to our national security because of the way the world has changed in so many ways; for example, you know innovation used to come from the government. The government would develop things like the internet, as an example, and then society would adopt it, and that's completely flipped and reversed today. Government can't move fast enough, and the innovation is overwhelmingly coming from the private sector.

Government struggles to keep up with that speed, the need for that cutting-edge technology for both economic and military applications. But higher education forms that critical link between the private sector and governments. It's a connective tissue between them that enables the government and others to leverage that technology, and I would argue that we have the best higher education system in the world. I mean all you have to do is look at the number of countries that send the leaders, that send their children to higher education in the United States, and the statistics are overwhelming. We have the best in the world, and that's really key that that partnership is there. Higher education educates government and military leaders as well as the entire private sector who will make the innovation to drive the 21st century.

So it means collaboration at all levels with the military and between the military and higher education—it's more important than ever. You know it's a dangerous world out there and on all our minds as we see what's happening in Ukraine and with Russia's illegal invasion, and I can't help but think that in the 21st century you would not think something like that could occur, but it is, and it demonstrates the importance of having a strong military, because there are folks out there who don't believe as we do, and these authoritarian regimes want to impose their will. So that increased collaboration will really help leaders in this area get to the innovation required to help us not go to war but to preserve the peace would be in my argument. As George Washington said in his first address to congress in 1790, he said: "to be prepared for war is one of the most effectual means of preserving peace." I think that's really important. To be prepared for war is one of the most effective means of preserving peace, and the consequences of failure can be devastating and mean even more conflict, what none of us wants, especially those who have seen the costs of war. So, the good news is again that there is a relationship between higher education and the military that's doing well—not that it can't get better, but it's doing well.

So, what connects that is the ROTC connection. I think it's one of the most important that is out there. ROTC is one of the first military higher education connections that most people think of, and it's really essential. The origins of ROTC go back to 1819, and a West Point graduate, Captain Alden Partridge founded the Military College mentioned earlier, Norwich. You have the consortium of military universities and colleges, and Norwich was founded and remains one of our nation's finest, as does the University of North Georgia. In 1916 the first official ROTC detachment was established at Harvard University and coincidentally the first superintendent of West Point, the United States Military Academy was Colonel Jonathan Williams, a graduate of Harvard University. So without American higher education, there would not have been a West Point, certainly not the success that they have attained indirectly through ROTC. And now ROTC programs are offered at more than 1700 colleges. Some of those are based out of the college that doesn't have a program, but they can attend a college nearby.

Army ROTC has 274 programs at colleges and universities, and in its first hundred years ROTC, from 1916 to 2016, more than one million officers were commissioned into the army through the Army ROTC. That's an incredible connection with our higher education out there across the country being involved in that. Army ROTC alone provides 274 million in scholarship money and more than 13,000 students each year; that's quite a connection. Navy ROTC has 77 programs at colleges and universities, and Air Force has 144. They're able to reach a total of 1700 schools because of campuses where they offer their program with nearby colleges. When you look just a couple of years ago, 56 percent of all newly commissioned active-duty officers in the Department of Defense were through ROTC commissioning programs. The largest source of commissioning is an extremely important link with society. Nothing against military academies or military-focused colleges and universities, but you need that broadening that comes from the connection with higher education across the country in that relationship.

Another way that there's a great connective tissue between higher education and the military is through military fellowships, and these are at all levels. You have some fellowships that are for more junior captains across the ranks, even enlisted NCOs and Sergeant Majors. There's also probably the largest, which are the more senior leaders, the Colonels (o6) level, roughly about 20 to 22 years of service, who are linked through the fellowship program, and there's 90 of those at the US army war college, and that goes across 49 locations and 55 different programs across the United States. Schools like Stanford, Duke, University of Texas, Columbia, and more—top universities and educational programs—and those fellows take advantage of a learning environment outside what is traditionally offered by the war college or military schools and military universities or army universities as mentioned earlier. Fellows experience these leading educational institutions, and it really helps prepare them for this highly complex and ambiguous environment. They act as ambassadors for the military out there and the majority being army ambassadors. They engage students, faculty, researchers, and the public, which is absolutely key.

Another critical link is through what's called the GI bill. The GI bill passed in 1944 to help World War II veterans, and what a difference that did make. It would free up tuition on that day by $500, which was enough. It was tremendously successful; almost 49% of college admissions in 1947 were through the GI bill. In 1956, 10 million veterans had received GI Bill benefits, and 10 million veterans of all ranks had received those benefits, and that had played a massive role in shaping the prosperity of post-World War II America. Many people have a lot written on this; they feel that was one of the absolute keys to America's incredible success post-World War II that enabled hundreds of thousand people to get an education who otherwise may not have been able to afford it. And now, of course, we have the modern version of the post-911 GI bill which was started in 2009, and the veteran's administration has provided educational benefits to 773,000 veterans and their family members. Since that 2009 post-911 GI bill with more than 20 billion dollars in benefits—and the modern post-9/11 GI bill is for all ranks, this is not just for offices but all ranks those who serve, officers, and enlisted can pursue an education, even while on active duty and certainly post active duty or in the guard in reserves as well. So it's one of the best examples of higher education and military collaboration at all levels.

Another is military research and development. I mentioned the military relies on the expertise of committed talented faculty members at our higher education institutions. In 2020 the Department of Defense contributed seven billion dollars to colleges and universities for research and development for the military. American university researchers help with defense-related research on cutting-edge technologies that will be vital on future battlefields and vital to keep us off of future battlefields. I would mention things like cyber, communications, biotechnology, artificial intelligence, and robotics just to name a few and there are many more. The research makes a real difference for our troops deployed overseas. I can speak from first-hand experience: in combat operations in Iraq and Afghanistan, key technologies were used to help with such things as countering improvised explosive devices that were tragically killing and maiming hundreds of soldiers. The army stood up a number of open campuses as an example where they partnered with higher education and the military and where army research lab scientists work with higher education. Five campuses are based at universities and spread out throughout the region. One example is the University of Texas at Austin where the army's future command, the command responsible for what things look like in the future and how we prepare for the future, is right there at the University of Texas Austin.

Higher education is also a partner in human relations, psychological studies, and strategy for the near term and future. So it's not just those technologies but also in other key areas. For example, in Iraq and in Afghanistan, researchers from American human terrain teams helped our troops in understanding the cultures in Iraq and Afghanistan. They accepted great risk to teach our soldiers more about local populations to learn more about that, and brave academic researchers like Michael Bachia and courageous women like Paula Lloyd gave their lives for America while serving in the combat zone from higher education. Just unbelievable dedication and sacrifice, and they will never be forgotten.

All those links are key to having the most professional military we can possibly have and therefore preserving the peace. American academia asks the tough questions that make our military better. Education helps broaden views and perspectives and helps eliminate silos, echo chambers, and group think that can occur. Without this in our military, we would have a failure. And failure means more conflict. Just from personal experience, I was fortunate enough to go to the University of Virginia early in my career as a young captain and get a master's in education, and I will tell you I used what I learned. It broadened me, and it enabled me to see other perspectives, and then it taught me. When I was at West Point, I was not thrilled with technology. The cadets there made me hate computers, but at the University of Virginia, I learned to leverage technology, and all that incredible broadening experience I used throughout 38 years of service, I can assure you.

There are also strong relationships in other areas, but those are the main ones, and it makes a big difference. However, challenges do remain, and let me cover some of those challenges. There are a lot of strengths,, but there are challenges, and we need to get better. First and foremost, just democracies as a whole and certainly in the United State,s tend to have an overly narrow view of war. Americans tend to view themselves as either at war or at peace, and that's what we've experienced, but that has changed in the 21st century, and we've seen it over time. The United States has entered a time that does not look like what most Americans think of as a war, but it certainly cannot be called a time of peace. Our adversaries see the world differently, as military theorist Carl Von Clausewitz said, "war is nothing more than the continuation of politics by other means." After the cold war, adversaries like China and Russia realized the overwhelming US military superiority. They knew that they would lose a conventional fight, but they did not just accept the US-led international system; they waged war through other means. Now while they built their military strength, our adversaries turned to what George Keenan described as the employment of all the means at a nation's command, just short of war, to achieve its national objectives. So, adversaries look for ways below the threshold of conflict to undermine the United States and our allies, and those are things like disinformation campaigns.

How much of that have we seen here? Recently things like cyber-attacks, these things that 20 years ago you didn't have to worry about as much, things like espionage, economic leverage, and intellectual property theft that involves businesses, the military, and our higher education institutions. So globalization has done some great things for the world connected us, but it also has presented some new challenges because we are so connected these things, images like disinformation campaigns, can move at a more rapid pace than ever and have a broader impact. So now we have a range of terms: hybrid warfare, gray zone warfare, and unrestricted warfare to describe this uncomfortable space that cannot be called conventional conflict but surely cannot be called peace. So it's vital that we recognize the blurring line between war and peace, and the competition between nations is not just about military equipment, and things that we typically think of but other areas. The significance it means the relationship between the military and higher education must be stronger than ever. No one entity is going to solve these problems, be it government, higher education, or industry. No one's going to solve these alone; this relationship's got to grow and be stronger to deal with this complexity and maintain peace.

Broadening our understanding of war is difficult because of a second challenge, and that's the growing military-civilian gap. In the United States on active duty, we have the smallest percentage of individuals serving in a peacetime era since between World War I and World War II, and it's really not a peacetime era. It's just incredible competition that's out there short of a full conflict. Post 9/11 wars largely did not impact the average civilian in any significant way. Fewer people today know someone on active duty, and military service tends to run in many families. I remember I was at an event a couple of years ago and about 300 general officers and senior leaders were there; they asked of those 300 how many have children serving in the military, and it was 99 percent. I myself have a daughter who's a Major, a military intelligence officer, and I have another daughter married to a major. It tends to run in the family, so what's happening is that gap is broadening. Especially after 9/11 military bases that used to be open and the public would be at the bases a lot, now it's more difficult, and they're closed up for security purposes, and there are fewer opportunities for civil-military interaction and less understanding.

So this lack of connection explains some of the mutual stereotypes and distrust that happened on both the military side and the higher education side. So the warrior class has shouldered the burden of our nation's wars for the past two decades, and they may feel like they will be unfairly criticized or judged in an academic environment, and some of the academics are wary of working with the military, whether due to moral or conflict of interest reasons. Challenges are compounded by the bureaucracy and the red tape. Unfortunately, the government is bureaucratic, and the military has an incredible bureaucracy. But whenever I get frustrated with military bureaucracy, I'd look at bureaucracy on the education side when my wife was earning her Ph.D. and feel pretty good about the military bureaucracy. So these bureaucracies are designed to say no and not necessarily work together in a manner that's needed for close cooperation. It's absolutely key, but it causes issues; there's a slowness that occurs with bureaucracy in both government and higher educational institutions. I look back to the example when we formed the Army University to link with universities to improve education across the army, going through the same accreditation process as any major university. For many years, we were trying to establish example partnerships with some universities and start up a Ph.D. in Strategic Leadership, as an example, and through the bureaucracy on the military side and the bureaucracy on the educational side, it literally took years, and unfortunately you have a change in folks, and it's so slow we never quite connected some of these ideas. So we've got to improve that and fight through the bureaucracy to truly partner better and stronger for the good of the nation.

Some efforts on the way, in the military, we just need to increase collaboration with academia in all ranks, enlisted and officer, from junior to senior. Again, it requires fighting against that bureaucracy to create new and innovative ways to make it work. One example was the senior sergeants major at the United States Army Sergeants Major Academy which used to go for a one-year program—a great course, but they got no credits for it. So they go for a year, and they got absolutely nothing. Now under Army University and the leveraging of the expertise, the Ph.D. has been developed and accredited so that senior Sergeants Majors, when they go for a year, they get a degree of various level depending on what they come in with. They can even go and get a master's degree at locations like Penn State because of that innovative partnership. We need more of those creative, new, and innovative ways to make it work. We need to expand the number of fellowships and ROTC. The military can never have enough junior and senior officers with a strong academic background who know how to challenge their own assumptions. That's something higher education does so well for us.

Military leaders should expand the emphasis on the value of the GI bill to the force. As I mentioned, we had the GI bill; unfortunately, only 40 percent of army veterans use that benefit. We've got to get better than that 40 percent using it. We need to expand that and what programs the military considers strategically important. For example, obviously, STEM fields are very important in the military and there's a pretty good effort in that, but also non-STEM related fields like area studies or humanities can also be critical to understanding the regions of the world our forces are deployed to and give our leaders a more nuanced understanding of those key things that are so important to understand when you're participating in any type event, be it from peacekeeping to humanitarian assistance, disaster response to conflict. We must remain on the cutting edge of STEM but also create that force that can think holistically about multifaceted issues for sure.

When you look at higher education, they must also fight through the bureaucracy to establish new partnerships. It's never easy, and it takes a lot of work, but with the right effort, I've seen that those schools have really put in the effort working together; there's incredible improvement, and I can give a lot of examples. One fairly close to where we are right now, when I was commanding the maneuver center at Fort Benning in Georgia, we had the idea of captains coming in to teach the courses at Fort Benning, our top-level captains in the military, to bring in the best to teach and instruct at Fort Benning. But they were missing out on the opportunity of a master. So we got with Georgia Tech University, and Georgia Tech actually came down to the maneuver center in Fort Benning so these captains could do, and they were very flexible and cut through the bureaucracy and establish a great master's degree program for those captains who were there teaching. So they didn't miss out on that broadening experience, and they were able to grow; those are the types of things we have to continue to work on and fight through some of the bureaucracy.

When you look at some of the red tape and bureaucracy that's out there, one of them is just credit for military courses and programs. As I mentioned, it's the same accreditation, but prevalent throughout higher education, they will not accept the credits from military schools such as through Army University for enlisted and officers, and part of this is against a bureaucracy but part of is also financial. They don't even put in the effort in some cases to accept the credits because they can get the GI bill and the individual can take more courses. But that also leads to fewer enrolled; that's why we're at 40 percent with the GI bill. One of the reasons for this is because there's frustration from the military that these credits aren't accepted. There's a happy medium where we can work it out so you can get better connection, better leverage with that GI bill, and give credits for those military courses that those veterans went through that are accredited at the exact same level of quality courses across the board. On the research side, we find that a lot of money is going into research in higher education to help the Department of Defense, but closer links are needed. Oftentimes, a university will compete for research; they'll earn that funding, and then they don't have a connection with the military. One example of this is when I was in the Maneuver Center, and we were working on training and improving training within the army, and a research grant went out to a university on the role of the brain in training and research, and it was several million dollars that they got. We got hold of them to ask, what are you doing in this area? Can we link with you to help improve training? They were grateful: goodness you called us; we didn't have any connection, and we didn't know exactly what the military wanted, and this really helps us, so we partnered. I find that that link is not there as well as it should be. Universities receiving grant money but not knowing exactly what's needed from them; it's got to be a collaborative process. It's much more impactful if they can work with the military, and even get soldiers involved, those who will be leveraging some of the products, technique. It's always better to get them involved early; the universities will benefit, and, obviously, the research will be higher quality; therefore, everybody benefits in the long run.

We have to work to develop a more open-minded perspective toward military personnel and academic circles. I talked about some of the military people who feel they might be judged by the academic community, well, it's the same in the academic community. Some will say, no, I'm not going to work with the military; I don't want to do; I don't want to promote war. I would argue that what you're doing is preserving peace through a strong military, and we have confidence in our nation that working together will help preserve the peace. How do we get at that? It's really key. Sometimes those stereotypes are out. For example, I was giving a leadership talk with an executive from Microsoft and Amazon at the University of Washington when I was commanding the first corps up in Fort Lewis and in Seattle. There were business MBA students in the crowd, and one of them said I have a question for the Amazon and Microsoft executives but not for General Brown because in the military you just tell people what to do and they automatically do it; you don't have to worry about leadership because they just automatically do it. I had to interrupt the individual and say, now wait a minute; think about what you just said. Leadership is leadership, and it's even more difficult when someone's life may be on the line. People are not robots; behavior is not automatic. These principles of leadership are even more important, even more they're even more intense when it's a difference between making money and someone's life is on the line. These are some of the stereotypes again. The military needs to do a better job of getting out, and higher education needs to do a better job by looking at how we can partner and ways we can do more together through all the many programs I talked about. By examining assumptions, more trust can be built between the military and academia; trust will enhance officers' reliance on quality academic sources to understand complex issues which are critical, because these issues are very complex.

A key way to achieving this is more engagements with veterans on campuses by encouraging the creation and supporting of veteran student centers and veteran organizations at higher education institutions, and this is a trend that is taking off. The Student Veterans of America, for example, has chapters on 1500 colleges and university campuses, which sounds great, but there are 4,000 colleges and universities. So, we have a way to go; 1500 is good, but we could reach 4,000 colleges and universities. I would encourage higher education to have these organizations advocate for veterans. Higher education is a great way to form connections between the military and the academic institution's service and educational events that can go a long way in closing that civil-military gap and eliminating unhelpful stereotypes from both sides. Lots of schools are also looking at this to attract more vets; as I mentioned, only 40% are leveraging their GI bills. That's a good thing; the more veterans that are attracted to the university, that better it is for the university, because they bring that GI bill money. But it's also good because they get a higher quality student, one who is more mature, who has gone through a lot, and who tends to be very successful at universities.

Let me conclude by saying that, because the world has changed, we can no longer afford stove pipes. We must deepen our partnership between higher education and the military. The only way to achieve this with the most professional military possible and preserve the peace is if we improve this and we continue to get better, get out of these stove pipes and collaborate, cooperate in working together. If we don't, we'll see tragic mistakes like we're seeing in other parts of the world now with just this incredibly myopic view that just asks, "how can this happen in the 21st century?" But it does, so you have to work on this. We must eliminate those unhelpful stereotypes that hinder this relationship. It's not military versus the academic or vice versa. It's not supporting war; it's preserving peace; it's a goal overall to preserve the peace that affords prosperity. Just these stereotypes are huge. To give you another example, when I was developing Army University, I had to brief retired four-star generals, and some of those were, like me, very old retired four-star generals, and they get set in their ways, and we had a course we were doing at the University of California Berkeley. We had numerous universities, but one was UC Berkeley, and as I mentioned that, several of the elderly four-stars backed off their chair. They were very upset claiming that's where protests began, and I saw that in the Vietnam War. So you can imagine their response when I informed them that not only are we sending captains there but we also have a professor from UC Berkeley coming to Army University to teach our mid-level leaders. I may not have done the best job describing it because I said, we need the expertise from the academy. We need this diverse viewpoint, and I challenged them. I said, I hope the professor shows up at the command general staff college with all the military folks there in their uniforms, and I hope he's got a ponytail, wearing Birkenstocks, shredded jeans, and probably the last thing I said is that I hope he's smoking a joint. Because the point I was making was we need that diverse viewpoint, different from that military mindset and higher education partnering. You can imagine these retired four-stars weren't all that thrilled about that, but I think they got the point. They understood that we've got to get out there and do this. The military and higher education are healthy, as I said. But I would give it a solid C plus, maybe a C that's average in today's world. With the complexity that's out there, we need to grow that to where that relationship between the military and higher education makes the dean's list, A-minus to B plus at the bare minimum, to be successful in today's world. Thanks very much.

2A

How Higher Education Fills the Security Gap in the Post-Cold War Era

Panelists:
- Dr. Craig Greathouse, Professor of Political Science and Associate Department Head
- Dr. Cristian Harris, M.A.I.A. Program Coordinator
- Dr. Edward Mienie, Executive Director, Strategic Studies Program & Partnerships and Associate Professor of Strategic & Security Studies
- Dr. Dlynn Armstrong-Williams, Department Head and Professor of Political Science
- Dr. Bryson Payne, Coordinator Student Cyber Programs

Moderator:
- Dr. Daniel S. Papp

As presented at the 2022 Strategic and Security Studies Symposium
Hosted by the Institute for Leadership and Strategic Studies
University of North Georgia

Dr. Daniel S. Papp:
It's a pleasure to be back up here in Dahlonega at the University of North Georgia once again. Welcome, everyone to the first panel at this year's UNG National Security Symposium. When our panel was created several months ago, it was then appropriately titled, "How higher education fills the security gap in the post-Cold War era." However, given the events of the last few weeks, I am taking the chair's prerogative and renaming and re-titling our panel to,

"How our education will fill the security gap in the new Cold-War era." We have five excellent presentations this morning, so in the interest of providing opportunities for the presentations and discussions, I've asked each presenter to make presentations in ten minutes or less. So, let's begin with our first presentation by Dr. Dlynn Armstrong-Williams, professor of political science and head of UNG's Department of Political Science and International Affairs. Dr. Dlynn Armstrong-Williams came to North Georgia in 1997 after completing her Ph.D. at Miami University in Oxford Ohio. Dr. Dlynn Armstrong-Williams was the founding director of UNG's Center for Global Engagement, with her research and teaching focused on international security, East Asian politics, comparative foreign policy, and international relations theory. She received UNG awards for both her scholarship and her teaching. Dr. Dlynn Armstrong-Williams's presentation this morning is entitled, "A state university's responsiveness to security education needs." The floor is yours.

Dr. Dlynn Armstrong-Williams:
[Please see peer-reviewed article entitled "Public Diplomacy and International Education" in this collection.]

Thank you everyone for joining me today. What I'm going to be sharing with you is sort of an example of what a state university can do to address these needs and fill the gap that we see, really, in security education. Just to give you a bit of context, Colonel Billy Wells is here. Dr. Billy Wells was my supervisor at one point during my career here at the University of North Georgia, and he was critical to the issues and the efforts that we're going to talk about. He was critical to the breaking down those silos that we and the opening speaker were talking about. We really began our efforts here in about 2007. The university was slightly different prior to its consolidation, and these numbers are from memory, so these numbers could be slightly off. I've been here a long time, and I can't remember what we had each year.

One of the things that you have to understand, I think, for securitization issues is, because as we saw, major concerns in the area of student preparedness are really in cross-cultural understanding and so linked to that, we initiated a campus-wide international plan. That campus-wide international plan could integrate many of the things that the original speaker discussed, including exchange opportunities, international internships, and a very strong curriculum. Students are prepared to really be global leaders and to build those bridges between higher education, business, and the military. As part of those efforts, what I'm going to talk about today is a little bit of this context.

We were trying to get kind of a finger on the pulse of what was happening in these desired transitions in higher education and security. But really, I think what provided that more national context was a HASC Report. The House Armed Services Committee provided this report, and I won't go into a lot of detail, but it talked about troop readiness and the fact that we were really confronting some issues in the wake of 911 and we saw that, really, we did not have a systematic approach to curricular change in making sure those who served in the military and security arenas were well versed in not only foreign cultures but also foreign languages, and we decided at this university that we were going to try to fill that gap.

You'll see from this report that our efforts really kind of gelled well with what was occurring at the national level. And I can't speak enough about this because this is one of the things we're struggling with now. We'll come back to that when I get to challenges. If you see the HASC Report—if you ever have a chance to see it—in 2008, it really talked about how we understood that we were a little bit behind in understanding regional expertise and languages and what at the time we called cross-culture skills. So as a university we decided and discussed how to respond. The Dean of Arts and Letters as well as the Deans of all our schools, myself, and Dr. Wells got together and talked about how we should best approach this. At the same time, Cadet Command for the US Army was also looking at these questions, so you'll see a synergy here that allowed us to push the envelope here at the university in lovely rural Georgia. I think that is the key that we expand these efforts beyond the beltway, and we expand these efforts so that we can really make this a national issue.

This is how we responded through curricular changes. We had a lot of things we thought we could do to really help soldiers and those that we were commissioning to become ready for this new reality, this more globally integrated reality. So, in 2007 we launched an International Affairs degree, and this degree was very successful initially. In the general courses you will see here, we wanted to make sure that students had a background in international relations and comparative politics because we were trying to give them that regional specialization, but also we wanted them to understand a global political context as well as global political economy context for all these major questions. Because of that, the degree allowed students to take these formative courses, then students had to choose a regional area of expertise; they couldn't be generalists.

We have four areas of expertise which are outlined on the slide, and students were required to take foreign languages up to the 2000 level, but we strongly encouraged 3000- and 4000-level language acquisition, particularly in strategic languages where we felt that there was a gap. So in this regional area of concentration, you'll see that students could take a history of that particular region and international studies focused on that region and upper-division language focused on that region. When I talked to incoming students, I'd tell them that they then have the opportunity to look through a microscope as well as a telescope. International relations is a telescope kind of vision of the international system, and comparative politics gives them the microscope vision. They can get down to the nitty-gritty cultural issues that are different in every country in their region as well as regional differences across the world.

We require an international internship in this degree. The international internship has to be within your region and that international internship has to have 60% cross-cultural interaction with individuals in the region. I was very blessed to have interned with the State Department as well as been a Fulbright Scholar. I had a lot of background in this, and what I found was that many students, while they spoke a language or they could handle themselves in an elite environment at university—which many universities across the world are—they very much struggle in day-to-day cultural interactions in another country. Because of that, this internship is designed to give them those day-to-day interactions with 60% of that time working with people in that country, and it was very important that we offered that. In the study abroad experiences, students have to spend at least 25 days in another country, but we encourage them to do a semester of study as that allows them to have those regional courses in the region which assists, and it gives them, of course, more cross-cultural competency because they're living in that country during those experiences.

We really encourage this and, as part of that, because I transitioned between the Political Science department and running essentially our International Education Office at the time, we were able to build strategic partnerships that served our students' needs. For example, we were able to build internships with the Asia Pacific Security Center because, of course, we have the Asia region of study, and we are very interested in security studies, so we built that partnership. We also have the ability to choose partnerships; for example, I am a Korean specialist, so we chose a partnership with Sogong University because they trained a variety of diplomats and governmental service people in Korea. We were able to choose partnerships strategically, which became very important to make them more viable for this degree. So, why did we get synergy?

This is what I want to talk about a little bit before I talk about the challenges.

At the same time in higher education, what you're going to hear in kind of the higher education nomenclature is high impact practices. While we were trying to do this with the support of the HASC, we also had a lot of transitions in higher education, nationally, and these high-impact practices are studying abroad, internships, and a capstone experience, all of which were integrated into the degree. While we had some momentum on the military side, we also had momentum on the higher education side. Linked to that we then were able to integrate a variety of high-impact practices into what we were doing.

So here are the challenges. The first speaker talked about these stove pipes; honestly, one of the first challenges we had was when Cadet Command narrowed their camp dates, which used to be spread across the entire summer. They narrowed the entire summer into the center of the summer. Before, we could send kids abroad before camp or after camp. That became very difficult, and those challenges then impacted time to the commission, which becomes another challenge that we have. That kind of transition of camp dates greatly impacted us.

The other issue we faced is, of course, international internships and studying abroad costs money. We were able to partner with Olmsted and other agencies. Many of our students get Gilman scholarships and Freeman Asia scholarships because they were studying strategic languages. Knowing these areas of the world, we've been able to increase our competitiveness for national scholarships as a department. In the past five or six years, we've gotten about 1.7 million dollars in scholarships for our students who are linked to this because they have this skill set. COVID of course has impacted us dramatically. It impacted us on this availability. Some of these high-impact practices we've had to modify, and that's been a huge challenge for us.

The other thing is that we have professional military education (PME) requirements and, hopefully, everyone in this room is somewhat familiar with those, so sometimes there are issues of communication because once the PME course are established, we don't want to change them because now they are already included into the 104R or plan of study which outlines their time to graduation. There have been issues in communication between professional military education requirements and what academic units are doing, which is part of that stovepipe issue. The other thing, of course, is that most of our cadets are contracted and because they're contracted, they have limited time to graduate. Now to be honest, what's given us flexibility is a lot of the cadets who join us tend to be highflyers. So, the cadets who choose these majors that are increasingly challenging tend to either CLEP out of courses or tend to dual enroll out of courses, which allows us more time in their plan of study to allow them to take advantage of these opportunities.

Where are we now? I wanted to show you a kind of graph for majors in my departments so you can also understand, I would say, other national challenges to what we're facing. You'll see that our international affairs degree, which is the orange line, peaked right in the center of the graph. Well, it peaked at a time period when the military, business leaders, etc., were all thinking about globalization, all talking about preparing leaders for this external world. We've seen the United States become increasingly inward turning, and as it becomes inward-turning then what occurs is there's less and less of this discussion out there. In fact, I remember I was confronted at one point and told that studying abroad was anti-American, and I said, "No;" the DOD is asking me to do this. They replied, "No, no; it's anti-American because Americans just need to stay here." So, we can't take this out of the global political context of the United States, I think, and we have to see how this impacts us.

You'll also see these numbers are impacted, and you'll see it nationally with a decline in enrollment and language education as well. We are struggling to try and make sure that we get individuals in the United States and elsewhere to understand the importance of this globalization effort for future US competitiveness and for strategic security reasons. This is where we need your help, and we need to have more and more people engaging in this conversation.

The other thing I want to point out to you is now that we feel that our students have this basis and cultural understanding knowledge, and they've spent time abroad and they've had these international experiences, what we think our next big challenge is sort of where we're going. One of the things that I'm seeing as the department head of political science and international affairs and what we will talk about a lot is cyber warfare and what's called the changing operational informational environment. One of the issues when we talk about escalating messages, right, is trying to reinforce messages that really serve our strategic interests. One of the questions is if you don't have a cultural understanding of particular regions, it is incredibly difficult to re-emphasize those messages properly. So one of the things that my department's looking at, and this is where I think this stove piping thing can come into play, is how do we do this? And how do we emphasize what we are really interested in, the values of the United States?

So because of that, we see more and more digital efforts by the Department of State; we see more and more digital efforts in diplomacy but not just due to COVID but because this is where the dialogues are taking place. So my department is looking at the possibility of integrating particular courses or skill sets that could help our students not only navigate that informational space but maybe start to adjust it themselves based on US strategic interests. So one of the things we're considering introducing here is we have to talk about bureaucracy in higher education. We here have to introduce courses of special topics to prove that we'll have the enrollment to prove that we have interests in those courses. That process can take a year and a half, then by the time I introduce the course, it then takes almost another year and a half to two years to get that course into the curriculum.

One of the things we're interested in is we already have a course in political polarity within the curriculum. We're really interested in looking at the role of information management in information manipulation and disinformation that enhances polarity around the world. We hope that through this course we'll be able to educate students about these efforts and also make them more aware of how to effectively assess information to understand if they are being manipulated in an international context. Essentially, we are facing a variety of challenges, but I think the answer to those challenges is an integrated curriculum that utilizes higher education as well as military folks, but we have to communicate about those threats clearly so that then we see where we can change the curriculum more effectively address those issues. Thank you.
[See Appendix for corresponding PowerPoint presentations.]

Dr. Daniel S. Papp:

Great, thank you Dr. Armstrong. If there are any questions for Dlynn or any of the other presenters, we'll handle that after all five of the presentations are made.

Our second presentation this morning is by Dr. Cristian Harris, professor of Political science here at the University of North Georgia and a member of the UNG's faculty since 2005, after having taught in Delaware, New Jersey, and Texas. Holding a Ph.D. from the University of Delaware, Dr. Harris is the program coordinator for UNG's Masters in International Affairs. He was awarded the university's highest recognition, the Distinguished Teaching Award, as well as other teaching awards. Dr. Harris's research interests include the role of US universities in public diplomacy and the political economy of Latin America. Dr. Harris's presentation is entitled, "International education and public diplomacy."

Dr. Cristian Harris:
[Please see peer-reviewed article entitled "Public Diplomacy and International Education" in this collection.]

Thank you. Let me start by saying that I'm going to face an upward moment. This is probably going to be a presentation that is not going to be followed by a question, whether this is going to be on the test or not. The US government and institutions of higher education have had a long and at times troubled relationship. As a reminder of that, 2022 is the 160-year anniversary of the Land-Grant College Act. My presentation today focuses on the international aspects of such a relationship. My references to the US government are concerned with the role of the federal government.

US government initiatives in international education form part of what we call public diplomacy. Public diplomacy is defined as the government's efforts to communicate directly with foreign publics in order to bring an understanding of its country's ideas, values, institutions, culture, and policies. The most visible manifestation of public diplomacy is cultural and academic exchanges, military exchange programs, foreign broadcasting services, cultural centers, libraries, language institutes, art exhibitions, and cultural performances. Public diplomacy initiatives found fertile ground in US universities. Freedom to pursue research, rigorous peer-review evaluation of scholarship, and the link between research and teaching develops into a vital component of US soft power. US universities both private and public emerge as the most globally-oriented universities in the world and, in turn, became the model for other institutions of higher education around the world. US universities attracted and retained talent from the various regions and disciplines they were studying, and those students and scholars chose to make their careers in US universities, giving the US an edge in its relations with other countries. Thus, public diplomacy initiatives facilitated the creation of US intellectual capital and became valuable instruments of US soft power.

How Higher Education Fills the Security Gap in the Post-Cold War Era

I will argue today that government policy has had a significant role in the formation of international affairs specialists in academia, think tanks, and even the government itself. Moreover, private foundations like the Ford Foundation, the Carnegie Corporation, and the Rockefeller Foundation provided additional assistance to support the work of scholars and helped build domestic consensus around the importance of international education and scholarship. During World War II, the US government helped train linguists and regional studies specialists. However, it was during the Cold War that internationalization became a national security priority. It resulted in the formation of a remarkable group of scholars and students who possess the linguistic skills, historical understandings, and analytical intellectual curiosity to engage with the world.

During the Cold War, the US government funded several now-iconic initiatives. The US Information and Educational Exchange Act of 1948 had the goal to communicate with foreign audiences and promote educational, cultural, and technical exchanges. It placed the voice of American broadcast under the State Department. The National Defense Education Act of 1958, which was triggered by the launching of the Sputnik satellite, expanded the federal assistance for regional studies and advanced language training and had the long-term consequence of boosting college enrolment. The Mutual Educational and Cultural Exchange Act of 1961, also known as the Fulbright-Hays Act, supported international academic and cultural exchanges. The Higher Education Act 1965 reinforced the role of colleges and universities and increased financial assistance to students at the post-secondary level. One of its legacies is the creation of the Pell Grant. The National Security Irrigation Act of 1991 had the goal of training a new generation of international affairs specialists for the post-Cold War era. The Warren Awards, consisting of scholarships and fellowships for undergraduate and graduate students, are intended to provide financial assistance for foreign language study in return for work in the federal government after completing the programs. It also consolidated other initiatives like Language Flagship and Project Go, known as the Project Global Officer.

In closing, the current changes to the commitment to international education and research pose a serious problem to the US standing in the world. This is happening at a critical time in the history of the US. The US is facing the rise of new challenges and challengers alongside traditional threats to its national security. Understanding and preparing to face these challenges and challengers requires a body of experts training in foreign languages, transnational studies, and international relations. United States universities are facing a competitive environment of their own, which is now defined by global forces. Shifting public priorities and a waning domestic consensus undermine the intellectual capital and power of the United States and risk weakening national readiness. Investing built too hard to acquire skills requires a long-term commitment. Free from political interference over the short-term supporting international education is likely to yield immediate results and risks creating impressions that academia and US universities cannot provide the answers. However, support for international education and research gives the US an edge on its relations with other countries given the uncertainty of where the future crisis will come from. Developing the intellectual capital of the US is a critical investment in its national security. Thank you. [See Appendix for corresponding PowerPoint presentations.]

Dr. Daniel S. Papp:

Thank you, Dr. Harris. Our next presentation this morning is by Dr. Craig Greathouse, professor of political science and associate department head here at UNG. Craig holds a Ph.D. in political science from Claremont Graduate school. Dr. Greathouse has published articles on European foreign policy, security, and defense policy, strategic culture, strategic thought, international relations theory, and cyber war. He also designs and runs simulations on the international system which he has presented to the National Defense University and other elements of the US military. He also helped create and is the advisor for UNG's Master of Arts in International Affairs. Dr. Greathouse's presentation is entitled, "Graduate education and national security: Expanding understanding."

Dr. Craig Greathouse:
Thank you. This presentation is going to give a brief overview of where we need to look in terms of graduate education in terms of national security, specifically national security. When we start to talk about graduate education, I want to have a common definition for everybody in the room. We're talking about, basically, education leading to a Master's or Doctorate post-Baccalaureate level. We're not talking about issues of certificates or skill-building; they don't build the critical thinking skills across the broad range that a Master's or a Ph.D. will. When we want to focus on building a skill set and broader thinking, we want to look at building those degrees at the Master's and Ph.D. levels, and then we're going to address the problems of what we're seeing now with the online scam schools and there are a lot, and some of the online education which I'll talk about here.

When we start to talk about the importance of graduate degrees, I try not to push kids towards Ph.D.'s unless they really want it. Ph.D.'s are about creating that cutting-edge knowledge, but you're also spending a huge amount of time. Your spending years having to learn; you're spending years having to build that level. There's also a cost and time commitment that has to go along with Ph.D,'s. You could have anywhere from seven to ten years into earning a Ph.D., and that maybe a problem before you get out and are productive. Whereas, if you're talking with a Master's degree, that takes two years or, if you're doing it part-time, three or four.

One of the things that I think we need to look at when we're talking about national security is looking more at the Master's degree because it builds on those basic skills from the Bachelor's. It takes what you're using and expands on that. You're not having to get into the area of cutting-edge scholarship and learning the newest methods and learning, you know, everything that's ever been written on a topic. You can get into a skill set; you expand your critical thinking; you expand your writing and analytical skills in going through this level if you're held to the standards. You're going to get a basic understanding of the methodology. No, you're not going to get the highest level, you know, statistical analysis and everything else, but you are going to gain an understanding of methodology, use of data, and basically creating integrated research projects. Those are critical ideas about being able to operate and process and present information. Going forward, you're also going to start to really learn how to use theoretical frameworks and other heuristic devices, which allow for very consistent analysis. It also allows you to understand where your biases are and if you're effectively using those theories' heuristic devices. So, when we're starting to talk about these Bachelor's degrees related to national security, that's what I want to focus on.

There are certain areas and degrees where you can directly connect, and it's easy to talk about. So international relations, security studies, and political science, specifically, when we talk about comparative politics, gives us the area studies and public policy issues, how you create and run policy, geography, and GIS systems, sociology, and anthropology; most people don't think about these, but understanding cultures are at their baseline. Okay, economics, the role of economics in the world matters. Computer science leads to the issues of cyber security. Then modern history, and I need to stress this: you know, understanding Europe in 1100 is nice, but understanding Europe in the 20th century leading into the 21st century is much more useful. So, there may be historians out there, but you know that it's modern. And then languages, understanding how the people communicate is an important element.

So, when you start to look at some of these, we can break these Master's degrees down in some of these areas. Do they give you a broad scope, or do they give you very narrow information? International relations by nature are designed to give you that, but that broader focus is to give you the reason why things happen. Economics can also give you that, but then look at some of the other degrees they provide, those specifics that you can't get any place else that economics and international relations are not going to give you. Strategic studies are very specific literature that you're going to get into; cyber security is a very specific literature. You know language and geography, so you need but you can't go one without the other, and what we've seen over the past years is you lose if you're not focusing on those big picture ones; you start to lose an understanding of what going on in the system.

For example, understanding why the state works and non-state actors, if you look at Russia and what they've done over the last, say, five to ten years. They're not engaged in cyber war, but they've got a lot of non-state actors that are connected to them that are directly engaged in that, so you need to understand that relationship. If you look at what's just happened where a lot of people are reacting to Ukraine currently and you had a lot of people come out and say, well, the US should put a no fly zone on. What's the impact if you do that? What's the international influence if you start to put US fighter pilots directly in conflict with Russian fighter pilots? Right, what happens?

So when we started to build the Master's in International Affairs—in 2009, it came online, I believe—we looked at the broader issues. And that's what we wanted to create: what does somebody need to know in order to address and become effective within the broader idea of being an analyst? So, some of our foundational courses are US foreign policy and the processes; IR theory; and the theory of comparative politics; so those theoretical frameworks to understand international political economy, because once I take the economic model and put it into politics, it's a very different player. International security, global governance, research methods, and then they have built a capstone, so you have to take and tie all of that together. And then we did put in some specific classes both region- and security-focused. When we built our masters, we tried to give you the broader focus, and then give you some expertise if you wanted it.

What we've sespecially after 9/11 is you get a lot of people pushing flavors of the moment. The focus on terrorism studies, post 9/11, everybody wanted to study terrorism, and they forgot some of the bigger problems. Terrorism is an element; it's an outcome of certain bigger things. Intelligence studies are important, but if you focus only on that, how do you create the context of understanding within that process. Security studies is the same thing; it's a narrow field. You've got to give them the broader picture beyond just the security studies. Language gives you some stuff, but it doesn't give you everything, so you have to be able to balance. And then we get to the problems of what we've seen with the new online scam universities.

There are a lot of good online programs, but there's a lot of junk out there that has taken advantage of a lot of people, and what you've got is they have very canned content, they have limited time. Most of their degrees say you can get a degree basically by doing all your classes in eight weeks. Limited reading, to give you one example of one university online I know, they limit the reading to a hundred pages a week. Okay, you just can't understand the field. They tend to be focused, predatory focused on VA military benefits; why? Because that's a money source. They tend to have predatory advertising; you know, advertising tends to be more expensive, and they tend to provide faculty who may not know what they're talking about. That's a significant problem if you're expecting to train people for these issues going forward.

So, when we talk about graduate education as an element to help with US national security, you don't need certificates, you don't need badging; you need the full development of degrees and to learn how to think, and you need the bigger development. I would argue you have to focus on the Master's level rather than the Ph.D. You're still going to need the Ph.D.'s, but the Ph.D. level is extremely difficult and, usually, about 5o% don't survive that process that start it. You need a combination of the big picture and specific programs, and then you've also got to deal with an environment where you've got limited degrees and bad actors out in the field. So, you're dealing with a very difficult environment, and how do you address that? And where do you address that?
[See Appendix for corresponding PowerPoint presentations.]

Dr. Daniel S. Papp:

Thank you, Craig; that was extremely enlightening, and I completely agree with almost 110 percent of what you said. Our next presentation this morning is by Dr. Eddie Mienie, professor of Strategic and Security Studies also here at the University of North Georgia. Eddie holds a Ph.D. in international conflict management from Kennesaw State University. Before entering the academic world, Dr. Mienie served as the Republic of South Africa's Deputy Ambassador to Switzerland and the Political Adviser to South Africa's ambassador to the United States. He also served in the South African military and saw active duty in the field during the Angolan Civil War. Dr. Mienie has also consulted for both Applied Software Technology and SAP Software Solutions. Dr. Mienie's presentation is entitled, "UNG's Strategic and Security Studies Program: Meeting a National Need."

Dr. Edward Mienie:

Thank you, Dan. I thought the presentation by General Brown was just spot on. It was a brilliant opening address to this conference.

And I was very happy to hear a number of factors that he touched on; we here at UNG are tracking with that, and we are implementing that. And I'd like to just revisit some of these factors that he had mentioned so that you can see further in my presentation how we are tracking with that and how we are implementing these factors that General Brown had addressed.

He spoke about how higher education and the US military needs a strong relationship with each other. As a senior military college, I can assure you that you'll see later on that we not only have a strong and existing relationship but also continue to build and make it even stronger. He mentioned that we are now in an era when war is being waged through other means, that we are no longer in a war or a peace situation, that there's a blurring of the line; in other words, we are facing today multifaceted threats to our national security interests, and here at UNG, under the leadership of particularly Dr. Billy Wells and the Dean of our College of Arts and Letters, birthed a degree program called Strategic and Security Studies. So, I'm going to speak to you from the perspective of directing the program.

UNG's effort in establishing international internships, reaching out to foreign military academies where we can place our students to expose them to a foreign culture where they can come back a more well-rounded student is well on track. We also here at UNG have visiting cadets from foreign military academies, whereas General Brown was saying ties and relationships that are cemented at this level at this early age in their career as cadets become eventually military generals within their own military. What a wonderful situation to have the option of picking up a phone and speaking informally with somebody whom you had met 30 years ago while studying at your institution, invaluable.

So, I would like to let you know that UNG's leadership in response to the multifaceted threats we face and the birthing of the Strategic and Security Studies Program is in alignment with

UNG's strategic plan. You'll see the why we addressed here, that is, how both the degree program meeting a national need for graduates trained in security affairs and with advanced abilities in critical languages as well as a study abroad experience is essential. So, in terms of our strategic plan at UNG, its objectives one and three, we are in sync with that as far as our program is concerned. Creating academic programs designed to prepare ROTC students to succeed in a globalized high-tech world: how do we implement this? We cultivate high-quality interdisciplinary programs with Strategic and Security Studies, programs that integrate innovative and emerging technologies including distance learning. Objective three of UNG's strategic plan is that UNG will become a leader in internationalized learning with an emphasis on globalization and the needs of an emerging and military workforce: how do we implement that? We ensure that the military education academic program meets the needs of the next generation of military officers and supports foundational competencies for effective leadership in complex and uncertain environments.

So, the intent of our program is preparing ROTC and civilian students for careers in public and private sector security with emphasis on military and international applications, consistent with nationally accepted trends and standards in the discipline. The program allows students to pursue the degree in an area that meets their interest and their ability. The option to pursue one of seven concentrations that we now have within the bachelor's degree program makes the program both comprehensive and flexible; it then focuses on issues of national security from the US government level, including military as well as civilian agencies and international security issues for private sector entities including businesses and NGOs.

So, what's our mission statement? To educate and prepare the next generation of the army and civilian leaders for national security, foreign policy, and corporate careers or to enter graduate programs. Through our goals to accomplish our mission, our program seeks to educate strategic thinkers to grasp complex international problems, identify strategic challenges for the United States, analyze historical precedence and geopolitical contexts, identify appropriate solutions to strategic or national security issues, and to promote creative interdisciplinary thinkers who can apply learned skills to the intricacies of national security strategy and international relations.

What types of jobs do our students move on to? So military service: we educate cadets to have the intellectual ability to understand and wrestle with a changing strategic climate. We train cadets to be professional, effective junior officers in the US Army. So, Defense, State Department, and Homeland Security are just a small number of federal agencies that would be interested in our graduates. For instance, the Defense Policy Board Advisory Committee is where our students can find themselves ultimately working. Regional Desks for the Undersecretary of Political Affairs at the Department of State, and another State Department area would be Regional Desks for the Assistant Secretary of State. For Political-Military Affairs Regional Desk for the Undersecretary for Arms Control and International Security Affairs, the Bureau of Intelligence and Research at the Department of State, the Bureau of International Security and Non-proliferation, and the Office of Intelligence and Analysis at the Department of Homeland Security.

What about the intelligence community? Intelligence is one of the concentrations that I mentioned, of which we have seven, so I've just mentioned three. Of course, we have 18 agencies within the intelligence community, including the CIA Operations Analysis, the Directorate for Science and Technology and (Digital) Innovation, and DIA Analysis Science and Technology. NSA: we have, as you may hear now after my next colleague, we have a Center for Academic Excellence in Cyber Defense Education at this university which is another big initiative from our site.

We have students currently who've accepted a position with the Secret Service. Army intelligence is another one; the students who are graduating now in three weeks' time will be entering Army Intelligence. We have a student who has accepted a position at the Georgia Bureau of Investigation. Civilian jobs, think tanks: these are thought tanks that our students can move on to after graduatin. Also, we encourage them, of course, to do a Master's, thinking about working at a think tank. But, anyway, I've listed a number over there.

There are federally funded research and development centers which would be another area of interest for our graduates, and then multinational corporations that are interested in enterprise risk management. If you take a global corporation like let's say Coca-Cola, they are dependent on third-world countries, and some of their raw materials are going into the drinks that we drink. If there's an impact on the supply chain, there needs to be an early warning system in place so that alternative arrangements could be made to obtain those materials and those resources elsewhere. So, it's the same thing with GE, Amazon, Microsoft, Lockheed Martin, oil and gas companies. Graduate schools where our students would be looking, some are listed there: Georgetown, Johns Hopkins, Missouri State Defense, Texas A&M, American University School, and the University of Texas LBJ's School of Public Affairs.

We have a common curriculum, and the courses that we offer there are American military history—these are compulsory courses; war and society, international relations theory, intro to strategic and security studies, comparative security issues, US national security policy, and military geography. There are student learning outcomes specific to each one of the seven concentrations that I've got listed there. Currently, we have 91 students in the major in the program; it's come down slightly, but that's a general trend nationwide, COVID having no small impact on that, and, of course, funding is another issue. But the concentration that's the most popular at the moment is intelligence, followed by military science, history, and then cyber security.

So, the delivery of the program is 21 hours of common curriculum, and we have 21 credit hours in the seven concentrations; preconcentration 21 hours, we have a study abroad requirement. You cannot graduate with your diploma unless you've done a study abroad; we're very intentional about that, or an international internship. We have six hours of electives, and the ROTC students can choose any one of those seven concentrations, whereas the military science concentration is not for our civilian students; and, of course, the method of instruction delivery is mainly in-class and online.

I've already addressed the study abroad, so we collaborate well with the Center for Global Engagement, which also falls under Dr. Billy Wells to get our students situated and identify an appropriate study abroad opportunity for them, and then, the student learning outcomes are set at the course level and the administration of the program. Again, the program falls under the auspices of the Department of Political Science and International Affairs, but we have stakeholders as well with each one of the seven concentrations, and we collaborate with the department heads of those concentrations. So, that's it. Thanks, Dan.

Dr. Daniel S. Papp:

Our final presentation this morning will be by Dr. Bryson Payne, Professor of Computer Science also here at the University of North Georgia. Dr. Bryson Payne holds a Ph.D. in Computer Science from Georgia State University and has been a member of UNG's faculty since 1998. He is the founding director of the university's Institute for Cyber Operations, which is an NSA and DHS Center for Academic Excellence and Cyber Defense. Dr. Payne has a host of publications focusing on computer education, computer security, computer use, and computer methodology.

Dr. Bryson Payne:
[Please see peer-reviewed article entitled "RASCLS vs Ransomware: A Counterintelligence Approach to Cybersecurity Education" in this collection.]

Thank you, Dr. Papp. It's my pleasure to be able to close out the panel presentations for this session with an applied example of how higher education can impact security, specifically national security. Through some interdisciplinary research that Dr. Mienie and I have been able to pursue over these past two years with some undergraduate students, we're working with a capstone student this semester in presenting some of this research over the summer. So, I'll share with you an applied framework that came out of this interdisciplinary effort. Two of my fellow panelists have already mentioned the programs that they were instrumental in developing under Dr. Wells's leadership the Center for Cyber Operations, which is now the Institute for Cyber Operations, is also due in great part to Dr. Wells's leadership.

I've been fortunate to be a part of that Center and Institute for the past six years, and we think that this is an example of the type of applied research that we can put into use in training not just our soldiers and employees but also our citizens, and this is an application of counterintelligence training to cyber security education, and we're going to use the MICE and RASCALS frameworks. I'm going to talk about the MICE and RASCALS frameworks: those of you who've worked in intelligence may have already seen one or both of these abbreviations. These acronyms are traditionally the five main triggers for converting a foreign asset into spies to go against not only their best interest but potentially the best interest of their country. Money, ideology, coercion or compromise, ego, extortion, and grievance we call this the MICE or MICE plus G framework. Those triggers are effective because of certain psychological components of just being a social human being. So, the RASCALS framework is a little more nuanced approach to motivation, the psychological tricks that cause us to fall for a money trigger or an ideology trigger. So, RASCALS stands for reciprocation, authority, scarcity, commitment or consistency, liking, and social proof.

So, what we're doing with these frameworks is applying them to cyber security training. We've all received training either from our schools or from our employers or our government organizations on resisting phishing. Has everybody gotten that training? All right, just making sure. We know that people are out to get our login credentials, our personal details, our organizational information, and even access to our systems. But we tend to forget those, or we tend to fail to be able to apply them behaviorally. When it's 5:30 on a Friday evening and you're trying to head home and you get an email that the CFO needs something or an officer has requested that you get this turned in before you leave and dog on it, we just do it; we get it done even though we may not have checked to see if that was an authentic request from the right individual. The same thing happens when we're traveling abroad and we get an alert that someone logged into our account, and we click that link and we give someone access to our account.

So, what we want to do is to be able to apply a counterintelligence framework, the same kind of training that we provide for our cadets or our students before they go on. On one of these studies abroad opportunities through the university, we make them aware that they may be exposed to intelligence operations while they're overseas, especially if they're traveling overseas in their uniform or for an official event in cyber and military science with another military institution overseas.

So, we've been warned against the dangers of social engineering. As we've seen though, in 85% of the breaches that get publicized, social engineering is often the way that an adversary gets inside your network or gets access to your systems. So, our training is not as effective as it could be. Social engineering is just deceiving an individual into doing something that's against their best interest or against the best interest of their organization or their nation. So just like our students have to become aware so also our cadets have to become aware before they travel overseas that they may be a target. We all need to understand that we may be a target every time we interact with our phones, whether it's to respond to a text message, whether it's to open a link or an email or even to answer a phone call. Has anyone gotten phone calls that were pretending to be your automotive manufacturer, the IRS, or whoever it might be? So, these tactics and techniques are not just limited to email; we think of cybe,r but they're coming at us from multinational criminal organizations, nation-state actors, and terrorist organizations in the form of text messages, phone calls, all of the different ways that technology has allowed us to become accessible.

So the first step in this training is helping our students and, eventually, our subjects in the research study understand that they are a target not just in the general sense that we are all a target but just about any time you open your email if you scroll down far enough or open your spam folder, you can find one of those emails asking you to click through telling you that you'll miss out on this opportunity or somebody has accessed your information. You better protect yourself from just a click or a spam message or spam phone call. So realize that you are an intentional target, that the individual, every individual is a target is key.

So, the goal of this framework is to help our students take this counter-intelligence-based cyber training more seriously and retain the information better. But perhaps more importantly, our hypothesis is that the participants in this counterintelligence training will be better able to respond behaviorally in the moment, even if they're leaving work at the end of the day, even if they're traveling overseas. We want our employees, our citizens, and our soldiers to be better equipped to respond to these threats.

So, I'll just hit a few of the main triggers you've all seen. These are in social engineering training or phishing training at work, I'm sure, or through school, but for money; money is the bait: do this to get money, or your money is in danger. This is the Bank of America telling you your login has been accessed from Beijing; click here if this wasn't you. Well, we can click through if we're not careful, and it's not that we don't have the knowledge, it's that if you're hungry, angry, lonely, or tired, leaving at the end of the day or traveling overseas in a different environment, in a hurry, under pressure, you still may fail even though you've had that training, even though you understand the dangers. We want people to be able to apply these lessons behaviorally.

So there was a great example recently where some public figures' Twitter accounts were hacked, and they were claiming to give it back; if you'll send me a bitcoin, I'll send you back two. How do people fall for these things? Well, unfortunately, lots of people do. For ideology, and we're not just talking about extremism or polarized ideology. It could just be, hey, you really support this cause or this belief, and here's a chance to give money. And we've seen scams like this recently to support Ukraine. Donate to Ukraine, there are lists of fake websites out there, organizations pretending to send money to support Ukraine. Coercion or compromise still works in every setting. Right, if you don't take care of this, if you don't send me some money, I'm going to release this information about you, release this video of you, release your email history.

All of the ransomware attacks out there right now fall easily under coercion and ego and can be used against us or turned against us, perhaps even more easily than the previous three. Even famous hackers like Kevin Mitnick, one of the most infamous hackers out there who now runs a security consulting company, was goaded into attacking a fellow hacker several years ago, and all it took was a little bit of social engineering through media. And then, of course, one of the more powerful triggers we're seeing lately is a grievance. There is actually ransomware for hire services where if you are upset with your former employer, you can provide your login credentials online and an entire criminal organization will go after your former employer trying to deploy ransomware and split the profits with you. There was a real-life example of this without ransomware in the middle where an employee was dismissed from the American College of Education and held the school's Google password, their Google account, the school's Google email, and all the other infrastructure services hostage asking for $200,000 payment, a ransom if you will.

So due to our short time, I won't go through every element here. But what we're doing beyond the MICE framework that you've probably seen in some format in your social engineering training, we're helping people understand that reciprocation, you know, if I give you a little bit, will you do one little thing for me? can often turn into an escalation of commitment. Authority can be used to leverage a position to make you do something that you normally would not do. Scarcity and limited time, only three spots left, you can join now; commitment consistency, well you did this, you might as well do this next step, that escalation of commitment, these are the psychological triggers, liking and social proof. They get us to buy the latest product, but they also unfortunately cause us to fall for the latest cyber scam.

So, in conclusion, we want to use these frameworks to educate employees, citizens, soldiers, and officers to spot social engineering and other cyber threats from that counterintelligence perspective, and we believe that these frameworks can also be used by employers to identify insider threats which is, of course, becoming an even bigger concern in the realm of cyber. And this counterintelligence approach begins with the realization that we are all targets. Thank you.

2B

PUBLIC DIPLOMACY AND INTERNATIONAL EDUCATION

Dlynn A. Williams and Cristian Harris
University of North Georgia

[underwent separate external peer review]

ABSTRACT

The US government and institutions of higher education have had a long relationship. This paper focuses on the international aspects of such a relationship and examines the nexus between US foreign policy and national strategy and US universities. US government initiatives in international education facilitated the creation of US intellectual capital and became valuable instruments of US soft power. During the Cold War, US foreign policy and national strategy relied greatly on the expertise of US universities. However, current changes to US commitment to international education pose a serious problem to its standing in the world. This is happening at a critical time in US history. The US is facing the rise of new challenges and challengers, alongside traditional threats to its national security. Given the uncertainty of where future crises will come from, developing the intellectual capital of the US is a critical investment in its national security.

Keywords: foreign policy, international education, national strategy, public diplomacy, soft power, universities

Public Diplomacy and International Education

The US government and institutions of higher education have had a long and, at times, troubled relationship (Coombs 1964; Altbach 1971; Chomsky 1997). Notably, July 2, 2022 is the one hundred and sixtieth-year anniversary of the Land Grant College Act (the Morrill Act). This was the first US government initiative in support of institutions of higher education (Thelin 2019). This paper focuses on the international aspects of such a relationship and examines the nexus among US foreign policy, national strategy, and US universities. In this paper, the references to the US government concern the role of the federal government.

US government initiatives in international education form part of Public Diplomacy. Public Diplomacy is a government's effort to communicate directly with foreign publics in order to bring an understanding of its country's ideas, values, institutions, culture, and its policies (Melissen 2005; Holmes & Rofe 2016, Kerr & Wiseman 2018). The most visible manifestations of Public Diplomacy are cultural and academic exchanges, student exchanges, military exchange programs, foreign broadcasting services, cultural centers and libraries, language institutes, art exhibitions, and cultural performances.

Public Diplomacy initiatives found fertile ground among institutions of higher education. Freedom to pursue research, rigorous peer-review evaluation of scholarship, and the link between research and teaching developed into a vital component of US soft power (Altbach & Peterson 2008; Cain 2021). US universities, both private and public, emerged as the most globally oriented universities in the world and US research institutions became the model for institutions of higher education around the world.

US universities attracted and retained talent from the very regions and disciplines they were studying. And, in turn, those students and scholars chose to make their careers in US universities, giving the US an edge on its relations with other countries (Choudaha 2018; Kerr 2020). Thus, the attractiveness of US higher education created a virtuous cycle, which reinforced US soft power.

Public Diplomacy initiatives facilitated the creation of US intellectual capital and became valuable instruments of US soft power. As Stephen Walt (2018) notes, academicians with expertise in political science and history played an important role in designing the post-war order. Moreover, government policy had a significant role in the formation of international affairs specialists in academia, think tanks, and the bureaucracy of the US government itself. At the same time, private foundations provided additional assistance to support the work of scholars and helped to build a consensus around the importance of international education and scholarship. This synergy was explicitly evident during World War Two and later during the Cold War. During World War Two, the US government helped train linguistic and regional studies specialists (King 2015). However, it was with the Cold War that internationalization became a national security priority. It resulted in the formation of a remarkable group of scholars and students who possessed the linguistic skills, historical understandings, and analytical and intellectual curiosity to engage with the world. During the Cold War, US foreign policy and national strategy relied greatly on the expertise of US universities.

This paper will argue that current changes to US commitment to international education pose a serious problem to its standing in the world. This is happening at a critical time in US history. The US is facing the rise of new challenges (e.g., climate change, transnational crime, cyberattacks, and economic and technological competition) alongside traditional threats to its national security. These challenges require a body of experts trained in foreign languages, transnational studies, and international relations. Shifting public priorities and a waning domestic consensus around international education initiatives undermine US intellectual capital and power—and risk weakening national readiness. Investing in building hard-to acquire skills requires a long-term commitment, free from political interference. Over the short term, supporting international education is unlikely to yield immediate results and risks creating the impression that academia and universities cannot provide the answers. However, the support for international education and research gives the US an edge on its relations with other countries. Given the uncertainty of where future crises will come from, developing the intellectual capital of the US is a critical investment in its national security.

This paper is organized in the following manner: the first section introduces key concepts, such as soft power and public diplomacy, and reviews their different manifestations. The next section is a discussion of the evolution of US public diplomacy in the postwar period. The third section of this paper analyzes the role of US universities in international education and public diplomacy. The final section is the conclusion of the article. It argues that international education is a valuable instrument of US power with critical implications for its foreign policy and national security.

Soft Power

The conceptualization of soft power is associated with the work of Joseph Nye. It is a term that enjoys a long intellectual evolution over more than two decades. Nye (2004) defines power as "the ability to influence the behavior of others to get the outcomes one wants" (p. 2). He argues, however, that there are three different ways to influence the behavior of others: coercion, payments, or coopting them to want what you want. It is the last form which he defines as soft power. That is "the ability to get what you want through attraction and persuasion" (Nye 2004, x). The soft power of a country arises from three resources: 1) its culture when it is attractive to others, 2) its political ideals and values when they are shared by others around the world, and 3) its policies, both domestic and foreign, when they are seen as legitimate and having moral authority (Nye 2004, 11). More importantly for this paper and commonly overlooked, Nye argues that non-state actors also have and can exercise soft power (Nye 2004; Melissen 2005).

Nye first proposed the term soft power in 1990. Originally, he referred to it as "indirect power" or "co-optive power behavior" that is "getting others to want what you want" (Nye 1990, 31). In contrast to direct or command power, countries relying on co-optive power are still able to achieve their desired outcomes because other countries want to achieve the same outcomes or because they have joined and agreed to a system of rules and ideas that produces such results. Command power can consist of the use of force ("sticks") or threats ("carrots") to get other countries to do what you want them to do. The distinction does not make military and economic power mutually exclusive. "Carrots" can include the use of military assets, such as military-to-military exchanges, transfers of military hardware, or joint military exercises. Similarly, the use of economic sanctions can be interpreted as an example of "sticks." Command power rests on coercion or inducement; co-optive power rests on the attractiveness of a country's ideas, culture, or leadership. Nye regarded co-optive power as a third dimension of power in addition to military and economic power (the two dimensions associated with command power).

Wishing to engage and rebut the contemporary declinist literature of the late 1990s, Nye proposed that observers had mistakenly perceived how US power was assessed. The world had changed and was still changing at the time of his writing in fundamental ways, rendering old metrics obsolete. According to Nye (1990), "Raw materials and heavy industry are less critical indices of economic power today than are information and professional and technical services" (8). These changes were affecting countries differently. Countries lacking the flexibility and ability to adapt to the different world reality were at a disadvantage, such as the Soviet Union and China; the US was not among them. The risk the US faced was not power loss or absolute decline, but the pursuit of the wrong strategies in the context of a changing nature of power.

Nye (2009) later cautioned against misinterpreting the uses of soft power and falling into the trap of triumphalism. In the post-Cold War period, the US, as the only superpower, created the impression among IR scholars and policy-makers that US commands would be followed not only because they had been proven right by history but also because others had no other choice but to follow. Nye (2009) argued that this ignored not only the strengths, but also the limits of US power. Soft power alone could not produce effective foreign policy. Instead, soft power must be understood as part of strategy, combining both hard and soft power which he called smart power (Armitage & Nye 2007). It requires "developing a deeper understanding of the role of soft power and developing a better balance of hard and soft power in our foreign policy" (Nye 2004, 147). Soft power could help a state create a favorable context for foreign policy and, thus, save a state a lot on carrot and sticks. However, soft power could not completely replace military and economic power because that is, hard power.

Soft power is not the opposite of hard power; it is a distinction not based on category but on degree both on the nature of the behavior and the tangibility of the resources. Traditional sources of power generally emphasized tangible aspects (e.g., military force, economic factors). Nye argued that intangible aspects also mattered. This is what he called "the second face of power" (Nye 2004). A country may be able to obtain what it wants because it can set the agenda of political choices and it can structure the situations in the international system to get others to change their preferences. One country will choose to want the same outcomes as another country because they admire the other country's values, want to emulate their example, or aspire to possess the same prosperity and openness.

Soft power is neither influence nor persuasion. The ability to establish preferences tends to be associated with intangible assets, such as an attractive culture, political values and institutions, and policies that are seen as legitimate or having moral authority. Soft power is attractive power and in terms of resources, soft power resources are the assets that produce such attraction (Nye 2004).

The discussion of the nature of power reveals that there is a historical discontinuity. In an era of trans-national interdependence, power is becoming "less transferable, less tangible, and less coercive" (Nye 1990, 33). Soft power, thus, has emerged as a key concept in international relations because world politics has been transformed in fundamental ways: the rise of the information age, the changing role and effectiveness of military force, and the inability to counter terrorism and extremism with hard power alone (Nye 2004). The art of persuasion to achieve political purposes is a necessary tool of political leadership in world politics. If a country can build consensus and portray its power as legitimate, it will face less resistance and enhance its power. If its political leadership and ideology are attractive, other countries are more likely to follow it.

The Instruments of US Soft Power

The instruments to build US soft power are public diplomacy, international broadcasting, student exchange programs, faculty and scholar exchange programs, development assistance, disaster relief, and military-to-military contacts (Nye 2004). To this list, we can add cultural centers and libraries, language institutes, art exhibitions, and cultural performances. These programs are not integrated into a single US government agency or comprehensive national security strategy. Instead, they are scattered across many departments and agencies of the US government and lack overarching coordination.

The challenges of wielding soft power are many. States do not necessarily control the content and delivery of culture. The sources and creators of soft power include not only government policies but also actors in civil society such as musicians, movie stars, professional athletes, business leaders, philanthropists, and of particular importance for the scope of this paper, scientists and universities. While working to create an enabling environment for its policies, government efforts may take years to produce the desired outcomes. This may prove particularly frustrating to governments wishing to cash in on their international goodwill. There is a gap between creating and using soft power. States must carefully cultivate international goodwill and capital today which can be called upon in the future. Moreover, the efforts mostly work indirectly by creating an enabling environment for foreign policy to advance.

Public Diplomacy

Public Diplomacy is one of the key instruments of soft power (Melissen 2005). Public diplomacy is a government's effort to communicate directly with foreign publics in order to convey an understanding of its country's ideas, values, institutions, culture, and its policies. Public diplomacy does not carry the negative connotations of propaganda. It is not a one-way form of communication conducted by diplomats. It is an effort to build interconnectedness at the level of civil society with individuals and non-governmental organizations. For the same reason, it is not a uniquely state activity; private actors can engage in public diplomacy. It presents a variety of private points of view, in addition to the official government position expressed by diplomats. Not surprisingly, the most visible tools of soft power are manifestations of public diplomacy: cultural and academic exchanges, student exchanges, foreign broadcasting services, cultural centers and libraries, language institutes, art exhibitions, and cultural performances. In recent years, the spread of social media (e.g., Facebook, Twitter, YouTube) and electronic platforms (e.g., cell phones, laptops, tablets) have dramatically transformed public diplomacy.

US Public Diplomacy Programs

The first US government effort in information dissemination abroad was the Committee on Public Information (CPI), also known as the Creel Committee, established by President Wilson in 1917. The intended goal was to counter German propaganda and influence in Central and South America during World War One. As another effort to counter German and Italian propaganda in Latin America, President Roosevelt established the Division of Cultural Relations under the Department of State in 1938 (Nye 2004; Arndt 2011; Danilov et al 2020). US agencies, in particular the Department of State, assisted in establishing several initiatives to reach out to foreign publics and expand educational opportunities and cultural exchanges overseas; for instance, the International Visitor Program of 1941 was created to bring international leaders to meet with their US counterparts. Years later, the Fulbright Program (1946) was launched. It became the US flagship exchange program and accentuated the role of academic exchange programs in public diplomacy.

During the Cold War, the US government funded several now iconic initiatives. The US Information and Educational Exchange Act or the Smith-Mundt Act of 1948 had the goal to communicate with foreign audiences and promote educational, cultural, and technical exchanges. It placed the Voice of America broadcast under the Department of State. The National Defense Education Act of 1958, triggered by the orbiting of the Sputnik satellite, expanded federal assistance and boosted enrollments in both regional studies and advanced language training (King 2015). The Mutual Educational and Cultural Exchange Act of 1961, also known as the Fulbright-Hays Act, supported international academic and cultural exchanges. The Higher Education Act of 1965 reinforced the role of colleges and universities and increased financial assistance to students at the post-secondary level. One of its most important legacies is the creation of the Pell Grants.

US public diplomacy is generally associated with the work of the US Information Agency (USIA). Created in 1953 by President Roosevelt and under the scope of the Smith-Mundt Act of 1948, USIA had its roots in World War Two efforts to counter enemy propaganda and present a favorable image of the US overseas. During the Cold War, the USIA became the lead agency for US public diplomacy. Foreign broadcasting services (for example, Voice of America, Radio Marti) made US culture more accessible to foreign audiences by communicating in their own language (Chatten et al 1999).

As part of the USIA functions, the US established a large number of American Centers (AC). These centers were open to foreign publics and provided a substantial physical presence in the middle of large cities overseas (CRS 2009). They are stand-alone US controlled facilities staffed by US embassy employees or US government contract workers. They are spaces in which the local populations can engage US speakers on a wide range of topics. They show American films, offer English classes, house libraries, provide space for public diplomacy events, and advise students on how to apply for higher education opportunities in the US. With the absorption of USIA into the Department of State and the Secure Embassy Construction and Counterterrorism Act of 1999 (SECCA), most of these centers were closed and their functions moved into the embassy compounds. Following such a move, they became known as Information Resource Centers (IRC). There were 87 IRCs worldwide in 2015 (ACPD 2015). IRCs were absorbed into American Centers. The Department of State operated 30 ACs in 2019 (US State Department 2019). The combined total for 2019 was less than 100 (US State Department 2019).

Binational centers (BNCs) also form part of the public diplomacy strategy of the Department of State. BNCs are private partnerships through agreements between the host government and the US and mostly operate in Central America, South America, and the Caribbean. However, while US embassies managed BNCs until the 1990s, their funding was dramatically reduced (GAO 2010). BNCs became self-sustaining not-for-profit organizations which generate fees mostly through English Language education. There are nearly 100 Binational Centers in 2019 (US State Department 2019). These centers can provide spaces for US cultural programming, but they have no permanent US staff and are essentially distinct from the functions of the Department of State.

American Corners (ACs) are the largest interaction point for public diplomacy efforts by the Department of State. There were 479 ACs world-wide in 2015 (ACDP 2015). ACs contain books, magazines, DVDs and other materials about the US. The Department of State provides financial assistance to sustain an AC, but they are managed by private entities—libraries, local universities, and municipal buildings—and remain in the hands of a foreign national who is not paid by or overseen by the Department of State (GAO 2010; US Senate 2009). In addition to "brick and mortar" locations, the Department of State also operates Facebook accounts, Twitter feeds and YouTube channels, Flickr sites, and active blogs (GAO 2010).

Citizen Diplomacy

Citizen diplomacy is one of the most valuable programs of public diplomacy (Bhandari & Belyavina 2011; Mueller 2009). It represents a unique public-private sector partnership. The idea of citizen diplomacy is deeply rooted in US democracy. It is based on the belief that all individuals have the power to make a positive difference by contributing directly to the cause of world peace and international understanding. For decades, non-governmental organizations and individuals have strived to build enduring relations with individuals of other countries. The primary goal of citizen diplomacy is to allow foreign visitors to meet directly with ordinary US citizens and acquire a more nuanced and unfiltered view of the population, its values, its ideas, and its culture. An additional goal is to expand the dialogue beyond the government official to a government official channel. Student exchange programs hosted by volunteers and private families—such as the Kennedy-Lugar Youth Exchange and Study (YES), the Future Leaders Exchange (FLEX), the National Security Language Initiative for Youth (NSLI-Y) and the Congress-Bundestag Youth Exchange (CBYX)—are important examples of public diplomacy initiatives involving citizen diplomacy (AFS n.d.). These programs offer foreign students the opportunity to study in the US, as well as US students the opportunity to study abroad. American Field Service or AFS has been a leader in such international education programs. Established during World War One, it was reorganized in 1946 as a secondary school student exchange program to help maintain and strengthen the mutual understanding and goodwill that had been established during their wartime humanitarian efforts (AFS 2014; Bhandari & Belyavina 2011). During the Cold War and particularly after 9/11, the US Congress funded student exchange programs through the Department of State. In addition, since its creation in 1961, the Peace Corps has sent thousands of volunteers as public diplomacy ambassadors of the US. As private citizens, these volunteers have engaged in solving developmental challenges around the world one project at a time, providing a favorable impression of US culture.

Cultural Diplomacy

Cultural diplomacy involves the power of persuasion and attraction and is one of the instruments of soft power. It is defined as the actions governments and non-state actors take to promote their cultural values and ideals to the societies of other countries. The essential idea behind cultural diplomacy is to bring together different cultures, enabling dialogue and mutual understanding. Cultural diplomacy is practiced by governments as well as civil society. Working in the middle of inter-national relationships, the cultural diplomat deals with ideas, values, science, art, thought, and minds (Arndt 2011).

> A cultural diplomat, as a servant of government, attempts to make such relations flow more smoothly and productively so as to minimize damage to national interests and maximize the possibility that elements of the interaction may grow into sustainable positive contributions to both or all participating nations (Arndt 2011: 7).

Cultural diplomacy as a form of soft power is at work when a government seeks to actively promote inter-cultural exchanges as a way to promote its national interests.

Public Diplomacy in the Post-Cold War Period

In the immediate post-Cold War era, the National Security Education Act of 1991 had the goal of training a new generation of international affairs specialists. It created the Boren awards (Boren scholarships and Boren fellowships) to provide financial assistance for foreign language study in return for work in federal government after completion of the program. It consolidated other initiatives like Language Flagship and Project Global Officer (GO).

The US government response to the terrorist attacks of 9/11 had a negative effect on the role of international education as a dimension of public diplomacy. While US study abroad programs or US-hosted exchange programs had not suffered the same fate of other public diplomacy programs in the post-Cold War period, security concerns following 9/11 severely curtailed the number of international students coming to the US. This is particularly troubling because recent empirical studies have confirmed that

US-hosted student exchange and military exchange programs play in role in spreading democratic values overseas (e.g., Atkinson 2010). Thus they serve as a key instrument of US soft power.

In the wake of the terrorist attacks of 9/11, US government officials and members of Congress, scholars, and think tank professionals vigorously argued that more effort, in terms of personnel and US dollars, needed to be allocated to public diplomacy programs. After years of budget cuts, reorganizations, and personnel reductions following the end of the Cold War, public diplomacy programs were at a low point when the 9/11 terrorist attacks took place. The US found itself in the situation of being unable to exercise its soft power and to counter and address US misperception overseas.

The Council on Foreign Relations (CFR) called for "a commitment to new foreign policy thinking and new structures" (Peterson 2002). The CFR argued that "addressing the image problem [of the United States] should be viewed as no less than a vital component of national security" (Peterson 2002). Similar calls for reform were made by the Congressional Research Service (CRS 2004; CRS 2009) and the General Accounting Office (GAO 2006; GAO 2010).

Official public diplomacy efforts had fallen short and radical voices espousing highly negative views of the US occupied the public diplomacy space the US had vacated. Consequently, the US image overseas deteriorated and the US position in the world turned increasingly vulnerable. As new challengers and challenges arise, failure to invest in public diplomacy as an effective tool of US soft power risks undermining US national security and foreign policy.

US Public Diplomacy and US Universities

US universities have played a foundational role in US public diplomacy. From the very beginning of official government efforts in public diplomacy, US universities have been present. When the Department of State established the Division of Cultural Relations in 1938, Secretary of State Cordell Hull and Under Secretary Sumner Wells announced to more than 100 academic and intellectual leaders that the US government could not commit more than 5% of the facilitation of cultural diplomacy (Arndt 2011). Universities would therefore need to use their own funds for public and cultural diplomacy efforts with very little governmental interference, but they would be assisted by embassies and their communications networks (Arndt 2011). Since 1938, universities have dramatically increased their international activities and influence. As the Cold War intensified, US foreign policy and national strategy relied even more on the expertise of professors and graduates of US universities. Experts with backgrounds in history and political science were joined, and sometimes replaced, by experts with backgrounds in economics and business.

Today, US universities are facing a competitive environment of their own framed by global forces (King 2015). In recognition of such a challenging landscape, many US universities have moved beyond the exchange of students and scholars to Comprehensive Internationalization (CI). CI is defined as "a commitment, confirmed through action, to infuse international and comparative perspectives throughout the teaching, research and service missions of higher education" (Hudzik 2011). While not all universities are at the same level of internationalization, universities have been responding to the same global pressures as other non-state actors. US institutions, in particular, have been grappling with the global market of higher education rather than just a domestic market. As student enrollment shrinks due to population shifts within the United States, many institutions are hoping to lure international students to not only make up for lost tuition revenue but also to assist in the internationalization of their campuses (Cantwell 2015; Sa & Sabzalieva 2018). As state funding for higher education in the US decreases, universities are increasingly tuition driven. They are responsive to market pressures and networked with other institutions of higher learning across the world. In fact, the decision of the WTO to recognize education as a global service business has prompted many governments around the world to invest in promoting the internationalization of their own domestic university systems (Foskett & Maringe 2010).

Universities have become one of the largest US exporters of services. Not surprisingly, government decisions affect them profoundly. For instance, policy changes following 9/11 and under the Trump administration had a negative impact on the number of international students seeking an education in the United States. This was further compounded by the travel restrictions imposed because of the COVID-19 pandemic. The total number of international students enrolled in US universities fell following 9/11 but eventually recovered to surpass pre-9/11 levels. More recently, numbers of new international students have fallen for five consecutive years during 2017-2021 (IIE 2021). Although the trend of US students studying abroad had continuously increased over the years, it fell drastically because of the COVID-19 pandemic (IIE 2021). Furthermore, enrolments in foreign language courses in US colleges and universities has fallen over the last decade (MLA 2019).

From a curricular perspective, US universities are embracing the idea that globally competent citizens are not only needed from their stakeholders in business, but also in global civil society. Universities within the US are focusing on increasing the cross-cultural skills of domestic students, in addition to attracting international students. In the area of curriculum expansion, the University of North Georgia (UNG) has created a series of robust programs in strategic languages, international affairs, and strategic and security studies (www.ung.edu). These programs, which incorporate international internships and study abroad into the curricula, are critical to getting students the skills needed to compete globally. Apart from developing a global curriculum, US universities are increasingly pressured to appeal to a more global clientele and are engaged in activities that will increase their attractiveness and raise their standing in global rankings. Major public universities, barely funded by tax dollars at all, are working on research agendas, import/export considerations and much more that will increase their international rankings and bring the best and brightest students to their university. In fact, many institutions have not just "waited" for these students to find them; they have gone out of the US and established campuses throughout the world. New York University (NYU) is a strong example, but there are many, many more. NYU's global network university is comprised of three comprehensive degree granting campuses in New York, Abu Dhabi and Shanghai (Sexton 2013; C-BERT 2020; Wilkins 2020). NYU is now servicing campuses that provide local student access to research centers, museums, businesses and non-profit organizations (Sexton 2013; Wilkins 2020). Other universities, such as Carnegie Mellon, have multiple degree granting locations throughout the world (Carnegie Mellon n.d.). These increasingly global universities are not alone. The Ohio State University has chosen to engage globally through "gateways." These "gateways are to provide operational support for faculty research/teaching and international partnerships, a portal for study abroad, a location for international student recruitment and alumni gatherings as well as a location for academic and corporate training and a way to partner with Ohio based businesses trying to expand internationally (OSU n.d.). Ohio State presently has gateways in Shanghai, China, Mumbai, India and Sao Paulo, Brazil (OSU n.d.). Ohio State is considering Turkey, sub-Saharan Africa and Europe as possible gateway locations in the future (OSU n.d.). Not only are global campus networks expanding but, international partnerships between universities are on the rise (Sutton 2013; Banks, Siebe-Herbig & Norton 2016; C-BERT 2020; Wilkins 2020).

Universities are engaging in public diplomacy efforts constantly throughout the world. This increased engagement within the arena of public diplomacy will be one of the great areas of curricular expansion in the future, particularly at the intersection of public diplomacy and the use of information as an instrument of power. In a recent report entitled "Teaching Public Diplomacy and the Information Instruments of Power in a Complex Information Environment: Maintaining a Competitive Edge", the US Advisory Commission on Public Diplomacy suggested a new scholarly discipline to study and teach influence and its role within statecraft. This discipline would further clarify the differences between influence operations and propaganda (Walker & Finley, 2020). Conversations within political science and security studies need to expand to take into account the operational information environment (OIE), as a domain of conflict separate from land, air, space and sea (Walker & Finley, 2020). The report advises institutions to create curricula that examine both the role of disinformation and influence in warfare (Walker & Finley, 2020). By incorporating the OIE into coursework, along with the skills emphasized in globalization, students will not only be able to navigate across cultures, they will become increasingly sensitive to the vulnerabilities of their own culture to disinformation and influence. Many US institutions now have the infrastructure and curriculum to surpass that which was once provided by the USIA American Centers. As argued by NAFSA (2008), the world's largest organization of international educators, more than a decade ago

> It is time, as a nation, to be purposeful about international education—to employ it consciously, in a coordinated manner as one of the tools in the national toolkit for engaging with the world in pursuit of the objectives that we share with the world's people.

Conclusion

It is time for such an engagement, but what is the way forward? How can universities project a larger presence in public diplomacy but keep the independence needed to maintain their research and academic freedoms? Public diplomacy is not a one-way communication, but rather a give and take between parties to develop a deeper understanding of each other's cultures. Public Diplomacy relies on people-to-people contacts, as well as government-initiated programs. Public diplomacy capitalizes on unique public-private sector partnerships and demonstrates the positive impact that individuals can make in international affairs. In addition, due the complexities of an ever-changing information environment, universities are on the front lines in guiding students how to filter through misinformation and to understand their own vulnerabilities to undue influence from external actors. Through student and scholar exchanges, US universities have already surpassed the original vision of the Fulbright program. US universities are continuously expanding the global dimension of their work. However, all of the global linkages in the world will be ineffective unless universities make significant strides in true comprehensive internationalization of education. Universities need to prepare both students and scholars in their roles of citizen diplomats. Individuals, who are not well trained, yet placed in authoritative positions, will reflect poorly upon the United States and its values. It is the responsibility of universities to prepare their students and faculty into this new role. This paper argues that universities are able to fill the void left by US public diplomacy initiatives and work in a comprehensive way to strengthen the US image in the world.

However, this strategy is not without significant risk. The diverging purposes of the US government and institutions of higher education predate the end of the Cold War. Changes in the relationship between US foreign policy and academia were already evident since the 1960s. The Vietnam War and the Watergate scandal produced a schism in the relationship between academicians and politicians. Moreover, coordination between so many institutions of higher education with vested economic interests is bound to lead to problems and lack of focus. Universities are primarily driven by the pursuit of prestige, revenue, and research. Thus, the challenge the US government faces is to incentivize professors, scholars, administrators, and students to become engaged with US public diplomacy. Reliance on private actors to help shape the image of the state and influence the opinions of foreign audiences about that state in strategic terms can be hampered by numerous factors. More importantly, these issues also raise the question of whether working with civil society actors enhances the state's soft power when its image, values, and political ideas are based on the promotion of a strong and independent civil society. In addition, universities that send students and faculty abroad which behave counter to US values will not enhance US soft power among the population. Despite these concerns, we should consider moving forward.

Current changes to US commitment to international education pose a serious problem to its standing in the world (Walker and Finley 2020; King 2015). This is happening at a critical time in US history. The US is facing the rise of new challenges (climate change, transnational crime, cyberattacks, and economic and technological competition) alongside traditional threats to its national security (competition with China, power politics with Russia, nuclear proliferation of North Korea and Iran). Understanding and preparing to face these challenges require a body of experts trained in foreign languages, transnational studies, international relations and informational sciences. Shifting public priorities and a waning domestic consensus around international education initiatives undermine US intellectual capital and power, and risk weakening national readiness. Investing in building hard-to acquire skills requires a long-term commitment, free from political interference.

Over the short term, supporting international education is unlikely to yield immediate results and risks creating the impression that academia and universities cannot provide the answers. However, the support for international education and research gives the US an edge on its relations with other countries. Given the uncertainty of where future crises will come from, developing the intellectual capital of the US is a critical investment in its national security.

REFERENCES

Altbach, P. G. (1971, May). Education and neocolonialism. *Teachers College Record, 72,* 543–558.

Altbach, P. G. & Peterson, P. M. (2008). Higher education as a projection of America's soft power. In *Soft power superpowers: Cultural and national assets of Japan and the United States.* Yasushi, W. & McConnell, D. L. Eds. First edition. Armonk, NY: M.E. Sharpe. 37-53.

American Field Services (AFS). (2014). *A brief history.* [brochure]

American Field Services (AFS). (n.d.). Public diplomacy initiatives. Retrieved from http://www.afsusa.org/about-afs/public-diplomacy-initiatives

Armitage, R. L. & Nye, J. S. (2007). CSIS Commission on smart power: a smarter, more secure America. Center for Strategic and International Studies, Washington, DC.

Arndt, R. (2011, November). *Cultural diplomacy and international understanding.* Paper presented at the Josef A. Mestenhauser lecture series on Internationalizing Higher Education at the University of Minnesota, Minneapolis, MN.

Atkinson, C. (2010). Does soft power matter? A comparative analysis of student exchange programs, 1980-2006. *Foreign Policy Analysis,* 6, 1-22.

Bhandari, R. & Belyavina, R. (2011). Evaluating and measuring the impact of citizen diplomacy: current status and future directions. New York, NY: Institute of International Education.

Cain, T. R. (2021). Academic freedom and its useful past. In *Challenges to academic freedom*. Hermanowicz, J. C. Ed. Baltimore, MD: Johns Hopkins University Press. 245-275.

Cantwell, B. (2015). Are international students cash cows? Examining the relationship between new international undergraduate enrollments and institutional revenue at public colleges and universities in the US. *Journal of International Students, 5*(4), 512–525.

Carnegie Mellon. (n.d.). Retrieved from http://cmu.edu.global/presence

Chatten, R., Herrmann, L., Markiw, T., Sullinger, F., & United States Information Agency. (1999). *The U.S. Information Agency: A commemoration: public diplomacy, looking back, looking forward*. Washington D.C.: United States Information Agency.

Chomsky, N. Ed. (1997). *The cold war and the university: Toward an intellectual history of the postwar years*. New York, NY: New Press.

Choudaha, R. (2018, April 18). A third wave of international student mobility: Global competitiveness and American higher education. *The Center for Studies in Higher Education, UC Berkeley*, Research & Occasional Paper Series 8.18. Retrieved from https://ssrn.com/abstract=3169282

Coombs, P. (1964). *The fourth dimension of foreign policy: Education and cultural affairs*. New York, NY: Harper and Row.

Cross-Border Education Research Team. (2020, November 20). C-BERT International campus listing. [Data originally collected by K. Kinser and J. E. Lane]. Retrieved from http://cbert.org/resources-data/intl-campus/

Danilov, A., Danilova, E. V., Shpakovskaya, M., & Gavrilă, A. (2020). Origins of the US public diplomacy. *Analele Universitatii din Craiova-Seria Istorie*, 38(2), 89-97. Retrieved from https://search.ebscohost.com/login.aspx?direct=true&AuthType=ip,shib&db=30h&AN=147464858&site=eds-

live&scope=site&custid=ns235470

Foskett, N. & Maringe, F. (2010). *Globalization and internationalization in higher education: Theoretical, strategic and management perspectives*. London: Bloomsbury Academic.

Hermanowicz, J. C. Ed. (2021). *Challenges to academic freedom*. Baltimore, MD: Johns Hopkins University Press.

Holmes, A. R. & Rofe, S. (2016). *Global diplomacy: Theories, types, and models*. Boulder, CO: Westview Press.

Hudzik, J. K. (2011). *Comprehensive internationalization: From concept to action*. Washington, DC: NAFSA.

Institute of International Education. (2021). *Open doors 2021: Report on international education exchange*. Retrieved from https://opendoorsdata.org/annual-release/international-students/

Kerr, P. & Wiseman, G. Eds. (2018). *Diplomacy in a globalizing world: Theories and practices*. Second edition. Oxford, UK & New York, NY: Oxford University Press.

Kerr, W. R. (2020, April 9). Global talent and U.S. immigration policy. *Harvard Business School Entrepreneurial Management Working Paper*, No. 20-107. Retrieved from https://ssrn.com/abstract=3581039

King, C. (2015). The decline of international studies. *Foreign Affairs*, 94(4), 88-98

Melissen, J. Ed. (2005). *The new public diplomacy: Soft power in international relations*. Basingstoke, UK and New York, NY: Palgrave Macmillan.

Looney, D. and N. Lusin, N. (2019). *Enrollments in languages other than English in US institutions of higher education, Summer 2016 and Fall 2016: Final report*. Modern Language Association of America. Retrieved from https://www.mla.org/content/download/110154/2406932/2016-Enrollments-Final-Report.pdf

Mueller, S. L. (2009). A half century of citizen diplomacy: A unique public-private sector partnership. *The Ambassadors Review*, 46-50.

NAFSA (2008). *International education: The neglected dimension of public diplomacy. Recommendations for the next president.* Retrieved from http://nafsa.org/uploadedtitles/NAFSAHome/Resource_Library_Assets/Public_Policy/public_diplomacy_2008.pdf

Nye, J. (1990). *Bound to lead: The changing nature of American power.* New York, NY: Basic Books.

Nye, J. (2002). Limits of American power. *Political Science Quarterly,* 117(4), 545-559.

Nye, J. (2004). *Soft power: The means to success in world politics.* Cambridge, MA: Public Affairs.

Nye, J. (2009). Get smart: Combining hard and soft power. *Foreign Affairs,* 88(4), 160-0_9.

Ohio State University. (n.d.). http://oia.osu.edu/gateways.overview.html

Patel, D. (2021). Go abroad, young American: The Biden administration must democratize international education. *Foreign Affairs,*

Peterson, P. (2002). Public diplomacy and the war on terrorism: A strategy for reform. *Foreign Affairs,* 81(5), 74-79.

Sa, C. M. & Sabzalieva, E. (2018). The politics of the great brain race: Public policy and international student recruitment in Australia, Canada, England and the USA. *Higher Education,* 75, 231–253.

Sexton, J. (2013, Fall). The promise of international education: Building a more just and elevated civil society. *IIE Networker,* 19-21.

Sutton, S. (2013, Fall) The growing world of collaborative internationalization: Taking partnerships to the next level. *IIE Networker,* 40-41.

Thelin, J. R. (2019). *A history of American higher education.* Third edition. Baltimore, MD: Johns Hopkins University Press.

United States. Congress. Senate. Committee on Foreign Relations.

(2009). *U.S. public diplomacy: time to get back in the game: a report to members of the Committee on Foreign Relations, United States Senate, One Hundred Eleventh Congress, first session, February 13, 2009*. Washington, DC: U.S. G.P.O.

United States. Congressional Research Service. (2004). *U.S. public diplomacy: Background and 9/11 commission recommendations.* Washington, DC: Congressional Research Service.

United States. Congressional Research Service. (2009). *U.S. public diplomacy: Background and current issues.* Washington, DC: Congressional Research Service.

United States. Department of State. Bureau of Educational and Cultural Affairs. (2019). *Office of American Spaces, FY 2019*. Retrieved from https://americanspaces.state.gov/wp-content/uploads/sites/292/LORES_AmSpaces2019Report.pdf

United States. General Accounting Office. (2010). *Engaging foreign audiences: Assessment of public diplomacy platforms could help improve state department plans to expand engagement.* Washington, DC: GAO.

United States. General Accounting Office. (2009). *Higher education: Approaches to attract and fund international students in the United States and abroad.* Washington, DC: GAO.

United States. General Accounting Office. (2006). *U.S. public diplomacy: State department efforts lack certain communication elements and face persistent challenges.* Washington, DC: GAO.

Walker, V. & Finley, S. Eds. (2020). *Teaching public diplomacy and the information instruments of power in a complex media environment. US Advisory Commission on Public Diplomacy (ACPD). Retrieved from https://www.state.gov/teaching-public-diplomacy-and-the-information-instruments-of-power-in-a-complex-information-environment-2020/*

Walt, S. (2018, February 20). America's IR schools are broken. *Foreign Policy*. Retrieved from https://foreignpolicy.com/2018/02/20/americans-ir-schools-are-broken-international-relations-foreign-policy/

Wilkins, S. (2020). Two decades of international branch campus development, 2000–2020: A review. *International Journal of Educational Management,* 35(1), 311-326.

Yasushi, W. & McConnell, D. L. Eds. (2008). *Soft power superpowers: Cultural and national assets of Japan and the United States.* First edition. Armonk, NY: M.E. Sharpe.

RASCLS vs Ransomware: A Counterintelligence Approach to Cybersecurity Education

Bryson R. Payne, Ph.D.
University of North Georgia
bryson.payne@ung.edu

Edward L. Mienie, Ph.D.
University of North Georgia
edward.mienie@ung.edu

Obulu J. Anetor
University of North Georgia
ojanet8061@ung.edu

[underwent separate external peer review]

Abstract

The past decade has seen an exponential increase in destructive malware attacks, including ransomware, in which victims' data and sensitive files are encrypted and held for ransom. A common factor in the spread of these malicious attacks is social engineering: the use of deception to manipulate individuals into divulging confidential information. Phishing emails are the most prevalent form of social engineering online: over 90% of computer-based attacks and breaches involve the element of phishing (Verizon, 2017), 74% of corporate cyber-espionage actions involve phishing (Jentzen, 2018), and 70% of breaches associated with nation-state or affiliated actors involve phishing (Jentzen, 2018). The authors propose a novel approach to cybersecurity education that explicitly incorporates counterintelligence principles to help inform users that their information is under attack and that their failure to identify a social engineering threat could lead to a data breach. The goal of the RASCLS vs. Ransomware (RvR) framework is that participants will take the counterespionage-based cyber training more seriously and retain the information better, and even more importantly, that participants in RvR training will be better able to respond *behaviorally* to fake phishing emails and related social engineering scenarios than participants trained via more traditional approaches.

Keywords: counterintelligence, cybersecurity, ransomware, social engineering

INTRODUCTION

There have been a number of high profile ransomware attacks during the past several years, against companies, individuals, hospitals, cities and states, military and government agencies. Newer double-extortion ransomware variants that first steal, then encrypt an organization's files have begun to make news since 2019. Double-extortion ransomware extends the concept of traditional ransomware—victims are first ransomed for decryption of their data, followed by a second ransom demand threatening to release the victim's stolen data publicly or on the dark web (Payne & Mienie, 2021). Research by Verizon (2017) suggests that over 90% of computer-based attacks and breaches involve the element of phishing, wherein legitimate users are tricked into giving their access credentials or other sensitive information to cybercriminals, cyberterrorists, or nation-state adversaries through deceptive email messages, text messages, phone calls and the like.

These human-centered attacks are known as social engineering. Social engineering aims to leverage human weaknesses to manipulate an individual into giving up sensitive information against their own best interest, their organization's best interest, or even that of their nation. Businesses spend in excess of an estimated $300 Billion (USD) per year on cybersecurity globally, up from as little as $3.5 Billion as recently as 2004 (Braue, 2021), yet cybercriminals and rogue nation-state actors use free email delivery to attempt to infiltrate millions of systems worldwide every day, including the most sensitive networks used in enterprise, government, military, and even intelligence agencies. Significant prior research has examined behavioral responses to anti-phishing education (Downs et al., 2007; Huang et al., 2009), but few training techniques seem to move the needle toward better behavioral outcomes.

Click-thru rates, the percentage of users who click a link in a simulated phishing email, are consistently in the double-digits across surveys (Kumaraguru et al., 2009), and there is little evidence that users are getting better over time (Terranova, 2020). In a 2009 study of 515 university users, 49.7% of users clicked through a phishing email before training, compared to 24.7% clicking through after training (Kumaraguru et al., 2009). In 2020, an international industry study spanning 98 countries found that almost 20% of trained users still clicked through malicious links in simulated phishing emails, with 13.4% of users actually submitting their username and password through the phishing link. Anecdotal experience among the authors of this paper corroborates these findings, with double-digit click-thru rates in multi-national corporate phishing training engagements.

After $300 Billion per year invested in cybersecurity, including billions spent on cybersecurity education and training programs for employees, soldiers, officers, contractors, and other users in sensitive and classified environments, to see 90% of computer-based attacks come through ubiquitous email seems discouraging. In fact, the very task of sufficiently training users to spot, prevent, and avoid phishing and other social engineering attacks seems daunting. However, we successfully train agents, officers, soldiers, diplomats, and other officials against human intelligence operations through counterintelligence education—otherwise, there would be countless more unsuspecting citizens turned into spies when traveling overseas or serving in high-ranking positions.

At the authors' institution, a public university built around a senior military college, students and cadets are trained in counterintelligence/counterespionage before traveling overseas for study abroad or international internships, and in many cases, those students returned with stories noting suspicious contacts or outright attempts to compromise them while overseas, but these attempts were thwarted by the training they had received. This marked difference between individuals being able to spot potential human intelligence techniques (spycraft) versus the abysmal statistics regarding human beings' ability to distinguish social engineering (phishing) attacks led to the present research.

In addition, insider threat often does not receive the attention it ought to, as it remains a serious security risk to a nation's national security interests. The paradox of underestimating the true nature of the insider threat and the damage a nation suffers after uncovering such acts of espionage ten and even twenty years after the fact can cause immeasurable damage to national security. In fact, in some instances, it may be impossible to gauge the true depth of the damage caused to national security interests.

A case in point is that of former CIA agent Aldrich Ames who spied for the Soviet Union for ten years undetected. As part of the highly sensitive sting operation to get him arrested, Ames surreptitiously received highly sensitive classified information to pass along to his Soviet handlers and he simply ignored that information. That was a major cause for concern and the question today that remains unanswered is what compromising classified information was he really after and for what purpose? Insider threat is a very difficult problem to get to grips with as the insider oftentimes has access to the most sensitive classified information. This paper proposes an approach to mitigating insider threat to the extent possible by having a framework against which insider behavior and potential social engineering can serve as an indicator to help detect nefarious activity.

The MICE+G and RASCLS Frameworks

Traditionally, the five main triggers for treachery are money, ideology, coercion or compromise, ego, and grievance (MICE+G). The most powerful of all may be the bitterness of grievance (Hughes-Wilson, 2017). In 1967, Navy Chief Warrant Officer John A. Walker decided to spy for the Russians purely for money (M) as he encountered financial problems in his personal life. In exchange for his access to all the code settings for the Navy's top-secret communications, he wanted money from the Russians. Walker received a sentence of life imprisonment. The so-called "Walker spy ring" is known as one of the most harmful espionage operations conducted against the US where the government had to pay around $1 Billion to have all the naval codes changed (Hughes-Wilson, 2017).

During the 1940s, Burgess, Maclean, Blunt, Philby, and Cairncross, spied for Stalin's Soviet Union, and became known as the "Cambridge Famous Five". These five hated their own fellow citizens and claimed to support The Communist International's interests that advocated world communism. These five penetrated the innermost secrets of the British state. Blunt spied on MI5. Philby ran the Iberian desk inside MI6, and Cairncross worked for Churchill's controller of secret services. Ideology (I) drove these five men to spy for the Russians against Britain and were eventually caught for their treachery (Hughes-Wilson, 2017).

Anna Montes spied for Cuba from 1985 to 2001, eventually working as a senior DIA analyst. She hated the US and idolized

Cuba. She was recruited as a spy for Cuba while working at the DOJ after being spotted for her anti-American outbursts and praise for Fidel Castro. She never spied for money and the mix of her own ego and ideology (E&I) motivated her to spy against the US. The FBI eventually led a sting operation against her and after she was convicted for espionage, she spent the next twenty-five years in prison (Hughes-Wilson, 2017).

Markus Wolf, head of the former East German state security (Stasi), was known for snaring potential spies by compromising (C) them with sexual indiscretions – the so-called honey trap. John Vassal was posted to the British Embassy in Moscow as a naval attaché assistant, when the KGB talent spotters identified him as a potential spy. After inviting him to dine at a restaurant in a private room, he was given enough alcohol to lose his inhibitions, and photographs were taken of him naked in compromising positions. These were used to blackmail Vassal. He acquiesced and exchanged his access to top-secret Royal Navy nuclear weapons material, including radar and tactics for the KGB's silence on his indiscretions. After a tip-off by the CIA to the British Admiralty, he was arrested, confessed to spying, and spent 10 of his 18 years in prison (Hughes-Wilson, 2017).

Egoist (E) Robert Hanssen was responsible for the FBI's worst security breach ever when he spied for the Russians before being caught 22 years later. He initially offered his services anonymously to spy for the Russians. After being overlooked for what he expected to be a promotion and he realized that his FBI career was not going anywhere, his ego got the better of him and he doubled his efforts to spy for the Russians for lots of money. He was eventually caught, found guilty of espionage, and sentenced to life imprisonment. Today he spends 23 hours every day in solitary confinement and all he has is himself and his ego to keep him company (Hughes-Wilson, 2017).

From 1961 to 1963, Colonel Oleg Penkovsky was the most senior Soviet officer to spy for the US and Great Britain. He decided to spy for the west because it grieved (G) him so much when the KGB chose to block his promotion to a higher rank. He grew disillusioned with the USSR and wanted to prevent a nuclear war between them and the US and he resented the way that Krushchev treated the military (Hughes-Wilson, 2017).

MICE may give us a superficial explanation for spying while not fully addressing the complexities of human motivation (Burkett, 2013). Intelligence case officers risk misreading their agents' behavior and motivations. Psychologist Dr. Robert Cialdini offers us six principles to help interact with others more smoothly and in a more beneficial manner represented by the acronym RASCLS. This framework may enable case officers to manipulate agents in a more subtle way. Using Cialdini's framework could offer case officers the opportunity to better understand their prospective recruits and increase their ability to identify other potential agents outside the MICE framework (Cialdini, 1984).

As part of the RASCLS, the "R" is for reciprocity. It is human nature to want to reciprocate in kind what they have received from another person. In American culture, this could be a business lunch, or drinks before getting down to business. This is done for the purposes of building rapport. Perhaps the potential agent has a small need that the case officer can assist with to build this rapport. This develops a feeling of obligation to reciprocate by the potential agent (Burkett, 2013).

The "A" represents authority and it is important to project authority during the recruitment phase. The case officer has the opportunity to show that he/she is part of a powerful organization with the backing of the US government. The best way to project this authority is for the case officer to refer to his/her knowledge of the overall operational environment (Burkett, 2013).

Scarcity, represented by the "S" refers to the belief that when something is less available, it is more valuable. A case officer will use scarcity to accelerate the agent's commitment to the project by conveying that their superiors need some kind of proof of their agent's unique talent to do the task at hand and get going (Burkett, 2013).

The "C" represents commitment and consistency as a main motivator for our actions. A case in point is the Korean War where Chinese interrogators managed to get US POWs to confess to war crimes on film where they "acknowledge" they used "germ bombs." This was followed up by written statements and declarations by these POWs and through this consistent messaging the Chinese managed to score some propaganda gains. Laying the ground for continued cooperation by showing how collaborating with each other can benefit both countries represented by the case officer and the agent fosters commitment between the case officer and the agent (Burkett, 2013).

People like "L" others who are like them often have a way of gravitating toward each other and cultivating deeper relationships with each other. By liking each other, the case officer is offered an opportunity to morph the initial, limited relationship with the agent into a friendship and eventually to a place where the agent believes the case officer is the only one who he/she can truly trust (Burkett, 2013).

The "S" represents social proof where people in unfamiliar environments observe others to determine the accepted behavior norm in that environment. This may explain why Oleg Penkovsky chose to continue spying for the west despite the KGB closing in on him. He was driven by social proof that the way of life as portrayed by the west was far superior to that of the USSR. Case officers reassure agents that others before them have successfully spied for the US and that they are doing the right thing to also spy for the US (Burkett, 2013).

MICE+G AND RASCLS APPLIED TO CYBERSECURITY

The same psychological tricks that intelligence officers use to recruit spies apply to social engineering in the cyber realm. Social engineering employs deception to manipulate individuals into giving up confidential information or performing other acts that are against their own best interest or the interests of their organization or nation. The focus of this research is to utilize counterintelligence training, based upon the MICE+G and RASCLS frameworks, to combat cyberattacks that are often initialized through social engineering. MICE+G and RASCLS are complementary frameworks, overlapping in a few dimensions, but the RASCLS framework is updated and aims to shed more light on the motivations behind espionage.

Perhaps the most fundamental counterintelligence concept to empower computer users, employees, and citizens against social engineering and related cyberattacks is to recognize that they are a target. Military officers and government agents are trained in counterintelligence (CI) techniques before traveling abroad to alert them to the dangers of being manipulated by adversary agents overseas. One of the first steps to this CI training is to make the subject aware that they are a target for foreign intelligence operatives. Based on this understanding, an individual is then trained in various tactics employed by foreign operatives in compromising potential assets and turning them into spies against their own best interest and that of their nation.

Similarly, the foundation of the RASCLS vs. Ransomware (RvR) training framework is to help individual information systems users that they are targets. Cybercriminals, terrorist organizations, and adversary nation-states are constantly searching for vulnerable individuals who can be duped into giving up their personal information, banking details, login credentials, trade secrets, and access to sensitive and/or classified data that may have value either now or in the future. Making individuals aware of the dangers of social engineering is the focus of many IT security training programs, but a counterintelligence approach begins with the realization not only that such dangers exist, but that every individual and organization is being targeted either directly or indirectly every time they go online, open an email, reply to a text message, or answer a telephone.

In the RvR training framework, users are first exposed to the MICE+G and RASCLS frameworks as a counterintelligence framework to combat espionage and given examples of some of the techniques in compromising soldiers, agents, and officials from the previous section. Then, users are presented with real-world cases of espionage involving individuals who were successfully manipulated with the MICE+G and/or RASCLS frameworks, with slightly expanded versions of the scenarios in the previous section, followed by specific training in the components of each framework, as follows:

Money – There are myriad examples to help train users to spot money as an enticement in phishing attacks, from easy money/found money/lottery notifications to loan scams, and newer bitcoin scams. Recently, several public figures' Twitter accounts were hacked or impersonated by cybercriminals claiming that a user could double their bitcoin accounts in the name of giving back to the community. The hackers claimed that if a user would send one or more bitcoins to an account advertised by the famous person's Twitter account but controlled by the hacker, the rich person would send back double that amount (see Figure 1). Needless to say, no cryptocurrency was returned.

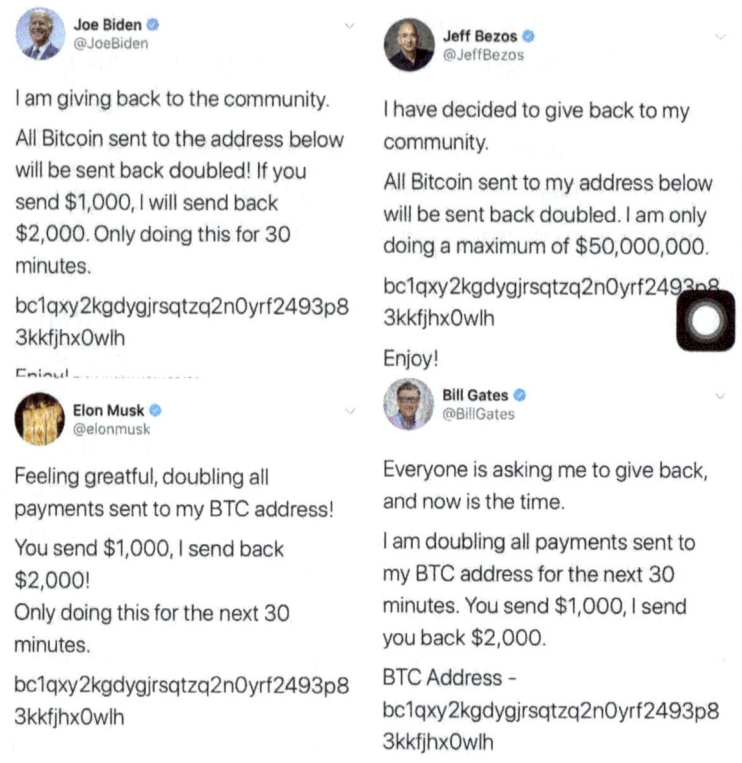

Figure 1. Money is used as the bait in this bitcoin scam delivered via compromised or spoofed Twitter accounts.

Ideology – Ideology can often be leveraged via Facebook or other social media disinformation, and not necessarily just extremist religious or extreme left- or right-leaning ideology. A recent example is the appearance of fraudulent Ukraine donation websites. Well-meaning individuals feel compelled to back up their beliefs by donating to what they see as a worthy cause (Figure 2).

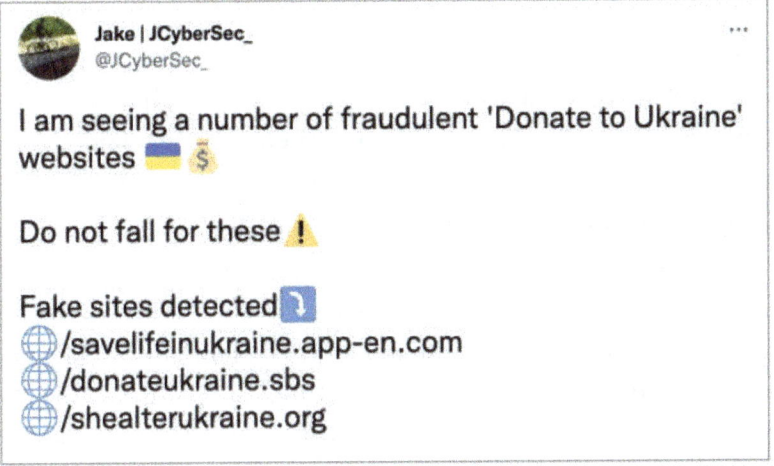

More fraudulent donation websites reported by *@JCyberSec_* (Twitter)

Figure 2. Ideology, even well-meaning, can be exploited to trick users into visiting malicious sites.

Coercion/Compromise – People often respond to danger or threats of blackmail. Double-extortion ransomware is a classic example of coercion—if a ransom is not paid, the attacker threatens to release the victim's sensitive data, including customer records, employee data, and more. Traditional, single-extortion ransomware is also an example of coercion, but demands a single ransom to restore a victim's encrypted files.

Ego/Extortion – Flattery or the need to save face can be used as bait to entice users to click through malicious links. One example of both ego and extortion is an email claiming to have captured webcam footage of the user naked, and demanding payment via bitcoin to prevent the footage from being emailed to the victim's entire contact list. The hacker claims to have installed a virus to hijack the user's webcam, but in fact, all the attacker has done is sent the threatening email.

Grievance – Disgruntled employees are a persistent insider threat. One dark web service offered to launch ransomware against a terminated employee's former employer as a "service", even offering to share the proceeds from the ransom with the aggrieved employee.

The MICE+G framework lays out the bait that many social engineering attacks use to ensnare their victims. The RASCLS framework exposes the psychological triggers that make users click through malicious emails in the first place.

Reciprocation – Though well-known, the so-called Nigerian prince emails still manage to trick new victims every year. The approach involves an unsolicited email claiming to be from a prince or other dignitary seeking to flee a country and transfer their assets, but needing help. The hacker offers a percentage to the victim if they'll help transfer money from one account to another. Sometimes the victim is asked to put up a small "good-faith" deposit of a few hundred to a few thousand dollars, other times a fraudulent wire transfer is made into the victim's account which they are asked to transfer or withdraw and deposit into a different account—but the wire transfer is reversed within the 72-hour window employed by banks to vet wire transfers, and the user has been tricked into overdrawing their account and sending their money to the hacker.

Authority – Among authority-based scams are those in which a caller or email claiming to be from a government's tax office, or immigration agency, or similar, demands immediate payment (often by gift card, cryptocurrency, or other untraceable funds) to avoid jail time or deportation. Or, an employee or soldier receives an email claiming to be from a superior and demanding immediate attention, from sending sensitive information to transferring money under the authority of the ranking officer or supervisor.

Scarcity – Claims of "limited-time only" or "just a few spots left" are so often used in marketing as well as online scams that it becomes difficult to differentiate between the two. However, during the financial crisis of 2008 when lending was scarce and loans were difficult to get, a group of hackers impersonated Bank of America, offering small, low-interest loans of a few thousand dollars to "loyal customers" needing debt consolidation services. Unsuspecting victims clicked through or called malicious toll-free numbers controlled by the scammers, resulting in theft of hundreds to thousands of dollars from some of the most vulnerable individuals with already-scarce financial resources and little recourse to attempt to recover any of the stolen funds.

Commitment/Consistency – Escalation of commitment is a well-researched psychological phenomenon in which an individual continues to invest time and/or resources precisely because they already have time and/or resources invested. Psychologists refer to this colloquially as "throwing out the baby with the bathwater", or throwing good money after bad. One fake scareware attack involves sending an email or popup message scaring the user into calling a toll-free number to receive help removing alleged malware that doesn't really exist. Victims provide their credit card information, and after they're charged, a number of other fraudulent charges are made, and the hacker convinces the victim they need to provide a different credit card number to pay for help stopping the attacks on the first account.

Liking – Dating sites are rife with scams and con artists seeking to defraud lonely individuals of money or sensitive information. Widows and widowers are at greater risk of dating fraud, with many losing their life savings to serial scammers posing as compatible matches through both reputable and disreputable online services. Similarly, affinity groups or websites may be used as a front to lead unsuspecting users to divulge sensitive information or credentials.

Social Proof – In marketing, this is termed the Bandwagon effect – everyone else is already doing something, leading an individual to take an action out of the fear of being left out. Cryptocurrency scams, claiming hundreds of thousands of investors have already profited, are an example of social proof used maliciously.

Recommendations and Conclusions

The MICE+G and RASCLS frameworks can be used to educate employees to spot social engineering and other cyber threats from a counterintelligence perspective, herein termed the RASCLS vs. Ransomware, or RvR, training framework. These frameworks can also be used by employers to identify insider threats from disgruntled or vulnerable employees. The focus of the training is two-fold, to help users recognize the attacks by understanding the motivating forces behind successful phishing attacks from a counterintelligence perspective first, and second, to modify the users' behavior by helping them use situational awareness to pause and reflect before responding to a phishing attempt.

A counterintelligence (CI) approach to cybersecurity education begins with the realization that every individual and organization is being targeted virtually every time they go online, open an email, reply to a text message, or answer a telephone. Whether cybercriminals, adversary nation-states, cyber terrorists, or simple con artists, malicious hackers can use email, text messages, phone calls, and more to launch millions of attacks per day. These attacks are both targeted at individuals working in high-value positions and broadcast to every user in an attempt to catch low-hanging fruit and easy prey.

The acronyms MICE+G and RASCLS are mnemonic devices to help users employ critical thinking *at the moment of attack* and to help users recognize cyberattacks via social engineering *before* they fall victim. More than mnemonics or traditional anti-phishing training, though, MICE+G and RASCLS are effective counterintelligence frameworks that can affect human behavior. This critical element, human behavior, is the key differentiator between the counterintelligence approach to cybersecurity training and standard knowledge-based training.

Future Work

This framework has been successfully deployed in courses at the authors' institution, but the authors intend to conduct an institutional review board (IRB) approved study comparing the effectiveness of the RvR counterintelligence training approach versus traditional phishing education as future work. It is hoped that the counterintelligence approach will better arm participants against malicious social engineering attacks, both short-term (through a quiz testing the users' ability to distinguish phishing emails from authentic emails) and over time (through a follow-up survey sent by email a month later).

More importantly than just recognizing phishing and other social engineering attacks, the counterintelligence approach is the same one used to help soldiers and agents be better able to respond *behaviorally* to espionage scenarios, and the goal of RvR training is to help users better respond to fake phishing emails and related social engineering scenarios than participants trained via more traditional approaches.

The authors seek to collaborate with faculty at other institutions and researchers in industry to weigh the effects of the RvR counterintelligence approach against more traditional, and even cutting-edge, social engineering training programs similar to larger-scale past studies like Purkait (2012) or Arachchilage, Love, and Beznosov (2016). Comparing the effectiveness of such instruction across industries is also an area of interest for future work.

References

Arachchilage, N. A. G., Love, S., & Beznosov, K. (2016). Phishing threat avoidance behaviour: An empirical investigation. *Computers in Human Behavior*, 60, 185-197.

Braue, D. (2021). Global Cybersecurity Spending To Exceed $1.75 Trillion From 2021-2025, *Cybercrime Magazine*. Sept.

10, 2021. https://cybersecurityventures.com/cybersecurity-spending-2021-2025/

Burkett, R. (2013). An Alternative Framework for Agent Recruitment: From MICE to RASCLS. *Studies in Intelligence Vol. 57, No. 1.* March 2013. https://www.cia.gov/resources/csi/studies-in-intelligence/volume-57-no-1/an-alternative-framework-for-agent-recruitment-from-mice-to-rascls/

Cialdini, R. (1984). *Influence: The Psychology of Persuasion.* Quill/William Morrow, 1984.

Downs, J. S., Holbrook, M., & Cranor, L. F. (2007, October). Behavioral response to phishing risk. In *Proceedings of the anti-phishing working groups 2nd annual eCrime researchers summit* (pp. 37-44). ACM.

Huang, H., Tan, J., & Liu, L. (2009, June). Countermeasure techniques for deceptive phishing attack. In *2009 International Conference on New Trends in Information and Service Science* (pp. 636-641). IEEE.

Hughes-Wilson, J. (2016). *The Secret State: A History of Intelligence and Espionage.* New York, NY: Pegasus Books, Ltd.

Jentzen, A. (2018). Phishing, Pretexting, and Data Breaches: Verizon's 2018 DBIR.

Kumaraguru, P., Cranshaw, J., Acquisti, A., Cranor, L., Hong, J., Blair, M. A., & Pham, T. (2009). School of phish: a real-world evaluation of anti-phishing training. In *Proceedings of the 5th Symposium on Usable Privacy and Security* (pp. 1-12).

Payne, B.R. & Mienie, E.L. (2021). Multiple-Extortion Ransomware: The Case for Active Cyber Threat Intelligence. *Proceedings of the 20th European Conference on Cyber Warfare and Security (ECCWS)*, 331-336, June 24-25, 2021, Chester, UK.

Purkait, S. (2012). Phishing counter measures and their effectiveness–literature review. *Information Management & Computer Security*, 20(5), 382-420.

Terranova. (2020). Gone Phishing Tournament: 2020 Phishing

Benchmark Global Report. Research technical paper. https://terranovasecurity.com/wp-content/uploads/2021/01/GPT-2020-Report-EN-1.pdf. Retrieved January 16, 2023.

Verizon. (2017). 2017 Data breach investigations report.

3

SYMPOSIUM TRANSCRIPT
COLONEL WORTZEL

Colonel Larry M. Wortzel

As presented at the 2022 Strategic and Security Studies Symposium
Hosted by the Institute for Leadership and Strategic Studies
University of North Georgia

I'm going to take a minute and just explain how what I'm going to say relates to what you're doing in college and, later, to the way the military now approaches dealing with foreign countries in general. It's what a foreign Area Officer does in the Army; that's what I was for a number of years, and I guess we still call it Human Terrain Analysis. Although, because of ethical questions by anthropologists, the Army no longer embeds anthropologists in units, as they stopped in 2015. In any case, this combines a knowledge of the geography of the area you're working in or studying, with cultural anthropology, the cultural and social dynamics, sociology, politics, economics, and religious beliefs. So, when you approach a country, you begin to appreciate the way its military services and its population thinks about questions, and that's what I hope to do with China. [Note: refers to accompanying slides.]

The first point I'd like to make is that the map of China is not always the size China was. It started as it was in the Qin dynasty in 221 BC and then between 220 AD and 290 AD, three of those central kingdoms went to war. By 940 AD between 111 BC and 940 AD, China had invaded and annexed Vietnam several times. During Mongol rule in China from 1231 to 1259 AD, they took over Korea, and the Yuan Dynasty, which succeeded the Mongols in 1271 AD, invaded Burma, Vietnam, Sakhalin, Laos, and Japan. You'll find the Chinese talking about themselves as a peace-loving country, as never being expansionist, but nothing could be further from the truth.

Zheng He was a Yuan Dynasty, General Admiral. He was a eunuch and was castrated at an early age and grew up in the court in the military between 1405 AD and 1433. He made seven expeditions and some of these had 300 ships and 27,000 men, artillery, infantry, archers, and cavalry. So, when he pulled up to a port and said, "would you like to trade," most of the small ports along the way said, "sure we'll be happy to," And it's kind of important. Xi Jinping's One Belt One Road route is Xi's principal project, and it's a sea route that really reconstructs Zheng He's voyages and is designed to get China's trade out. The significance of it from a military standpoint is that beginning, I'd say, in 2008 and 2009, the previous Communist Party Chairman told the PLA you've got to be prepared to be a more expeditionary force and to be able to protect China's interests along this maritime silk road. The old silk road is really the belt that goes all the way through the desert and Asia through Iran and was a major trading route down to India and even into Africa.

The other thing to keep in mind when you're dealing with China is that you're really dealing with a country, an army, a military, and a people who have been taught to have a giant chip on their shoulders. All the educational systems tell them that they went through a century of humiliation, but they don't include the fact that the reason they went through the century of humiliation is the Ming Dynasty was horribly corrupt and wasted its treasure and money and did nothing to govern the country. In 1839 to 1842 the Chinese lost the Opium War to the British, and that cost them Hong Kong and extraterritoriality in a number of places. In 1858 to 1860, there was a second Opium War where the French and the British invaded through Tianjin to Beijing and burned the emperor's summer palace. If you've been to Beijing, the old one still has ruins next to Ching Hua University and the new one is really rebuilt and has a marble boat in there that the empress built and used the money that was designed to create a fleet. From 1884 to 1885, the French fighting over access to the Red River on the Sino-Vietnamese border invaded China and sunk the entire southern fleet of the Chinese. In 1894 and 1895, Japan invaded, and China lost the whole Korean peninsula and Taiwan and most of north China, but western powers convinced the Japanese to withdraw, although they kept Taiwan and Korea.

The Boxer Rebellion in 1900, a rebellion among Taoists in China, rose up and killed a lot of westerners. So, eight countries including the US, England, Germany, and Japan, invaded China again from Tianjin to Beijing and burned the summer palace and the imperial city again. And I'm proud to say I was a member of the ninth infantry which was part of the invasion force and took the entire silver treasury of the emperor and melted it into a big punch bowl which sits in the ninth infantry headquarters today. And if you're an officer in the ninth infantry, you toast with one of those original silver goblets. In 1931, Japan invaded again and took all of northeast China turning it into Manchu. And then in 1937 to 1945, of course, World War II occurred, which interrupted a civil war in China.

In 1985, Liu Huaqing, who was an Army general who fought the Japanese and World War II and fought the nationalists, was made the third commander of the Chinese Navy. He came up with this vision for the Chinese Navy between 1985 and 2040, and they were to be able to exercise sea control and air control with the Naval Air Force and Air Force over the first island chain. It encompasses the South China Sea and the East China Sea, which are claimed by Japan, Korea, and Taiwan, and really claims a lot of the islands that are also claimed by Vietnam and the Philippines, and some now by Indonesia and Malaysia, and Singapore's not involved there. He wanted to seize control there, and over the long term, he wanted an expeditionary Navy, a Blue-Water Navy, and Air Force that could travel outside the first island chain and give him a sea denial capability out to 1500 to 2000 miles to the second island chain. You see how important and critical Japan and the US-Japan alliance are. You see now where Australia comes into play because it's just south of Indonesia, and they've done quite a bit, if you've followed their naval construction and their exercises, to actually accomplish this.

The ranges they're trying to get coincide with the combat radius of a US aircraft carrier or the range of a tomahawk missile. The idea is, in the event of a contingency involving Japan, the East China Sea, Taiwan, or the South China Sea, they want to keep US forces at bay and outside the tomahawk missile range. Meanwhile, they've developed ballistic missiles that through a combination of long-range, ground radars, sea radars, and space radars, and space collection supposedly can hit moving aircraft carrier battle groups.

It's an effective strategy; it's not completely in place, but what I'm going to go into next for a few minutes is what I would call competing strategies between China and the US. In the major straits between the first island chain and the second island chain going all the way down through Indonesia and Australia, the Chinese strategy depends on penetrating that first island chain before the US or any other countries can react, creating sea denial in the first island chain and or sea denial in the second island chain. The Japanese Army, the US Army, and the US Marine Corps with the Air Force and Navy have come up with really similar strategies to deny the Chinese military the ability to get outside the first island chain and really defeat their forces in detail before they can penetrate the first island chain. For the Army, it's called the Multi-Domain Task Force, and for the Marine Corps, it's Expeditionary Advanced Base Operations. The strategies are similar. Up and down this strait, there are US allies or partners, and that includes in some cases Vietnam as a potential partner, and the idea is you create multi-disciplinary task forces of artillery, cruise missiles, ballistic missiles, infantry, and in some cases mining, and you occupy these straits and channels quickly to deny the Chinese the opportunity to penetrate the first island chain out to the second island chain, and then you defeat them in detail.

So it's a pretty hostile and competitive set of strategies, and I've played dozens of war games on this, and, at times, they've gone to nuclear weapons, but very often they've gone to long-range ballistic missiles and cruise missiles strikes on the mainland of China, strikes on the mainland in the United States, and it's pretty escalatory. We have practiced it, I think we've refined it, and we work pretty closely with the Japanese on it. It really depends on the ability to get these forces out quickly and the cooperation of countries all along this strait, and those countries are really not happy being put in the position to have to make a choice because they have significant trade with China, although the Chinese have significant claims on some of their territory. Needless to say problematic, it's problematic; it would lead to an exchange of forces that in war games is highly escalatory, and we're dealing with two nuclear powers. Thank you.

[See Appendix for corresponding PowerPoint presentations.]

4A

RETHINKING HIGHER EDUCATION PRACTICES TO STIMULATE INNOVATION AND GLOBAL SECURITY

Panelists:
- Professor Jacek Dworzecki and Associate Professor Izabela Nowicka, University of Land Forces, Wroclaw, Poland
- Shannon Vaughn, Virtru Federal
- Iyonka Strawn-Valcy, Georgia Institute of Technology Director of Global Operations
- Magdalena Bogacz, Ed.D., Assistant Professor of Leadership and Ethics at Air University's Global College
- Crystal Shelnutt, Ed.D., Senior Lecturer at the University College of West Georgia

Moderator:
- Michael Lanford, Ph.D.

As presented at the 2022 Strategic and Security Studies Symposium
Hosted by the Institute for Leadership and Strategic Studies
University of North Georgia

Dr. Keith Antonia:

I want to introduce our second panel, Rethinking Higher Education Practices to Stimulate

Innovation in Global Security, and I'd like to introduce our moderator. Our moderator is Dr. Michael Lanford who is an assistant professor of higher education here at the University of North Georgia. His teaching and research explores the social dimensions of education with specific attention to equity, globalization, institutional innovation, organizational culture, and qualitative methods. In April this month, his first book entitled *Creating a Culture of Mindful Innovation in Higher Education* will be published by the SUNY Press, State University of New York Press. Additionally, Dr. Lanford has written approximately 3o articles and book chapters for such scholarly publications as the *American Educational Research Journal*, *Educational Philosophy and Theory*, *Higher Education Handbook of Theory and Research*, and *Qualitative Inquiry*. He has received funding to present his research in Canada, Hong Kong, Mexico, Taiwan, the United Kingdom, and the United States and has been a personal mentor to me as well. So, without further ado, Dr. Lanford.

Dr. Michael Lanford:
Thank you so much Dr. Antonia. I am excited about this panel. It's one that I've wanted to do for quite some time, talking through hopefully re-envisioning ways to create effective and innovative higher education programs that can support both society and the military, and so, without going into any greater detail, well I'll just introduce our first speaker today: Shannon Vaughn. He's had nearly two decades of experience with the US Army and Intelligence, has also served as a military diplomat and attaché, senior China cyber analyst, IT systems analyst and is a specialist in Chinese Mandarin linguistics. He's also currently the general manager of Virtu Federal, a data security encryption company that partners with government organizations, and I'm excited to hear his remarks and leading off this panel today so, Shannon.

Shannon Vaughn:
Thank you for having me. I appreciate it. When they asked me to do this, I wrote my abstract and I kind of said, "Why my story; is it a kind of an anomaly?" I think Billy Wells will tell you that I was his guinea pig here. I was a young cadet and, after three years,

right before I'm about to commission, I ended up finding out about this DLI program. Go down the road, and here I am, but I started thinking back on how did that come about? And I kind of put it into to two buckets, and I kind of want to talk about two buckets today: the first one is that hard problems take time; they just do. If you want to be an expert, you got to spend time with it. I think when we saw Larry Wortzel earlier he was a attaché from '88 to '90 in Beijing and he's finally to that realm where everybody in government will tell you Larry Wortzel's an expert on Chinese policy.

The second one I want to talk about is kind of how and is probably a little controversial, but the military doesn't reward kind of or have cultural experts, and I think it ties back to the first thing: hard problems take time. So we were talking about this last night at the at the icebreaker dinner, and I studied physics here and Spanish and Chinese, but I would never give a 12-year-old a Physics book and expect them to acquire what they need. But you can give a young kid a new language, and they're actually going to pick it up. That language acquisition is actually faster than when I started Chinese at DLI when I was 21 years old. But I think the other part of that is opening the aperture, getting young people to see what is out there.

I fortunately have four great siblings adopted from China, and so I was exposed to China from the time I was 12 years old, so part of our family traditions were traditional Chinese traditions. Even before I ever took the language, I had to spend time with the culture, and I think that that is a very key point, that when we talk about reimagining education for national security, getting people access to the language or the culture of the people, getting people overseas young is a very important way to do it.

As I said earlier, hard problems take time. Starting with Chinese at 12 was probably even too late personally, but it was a great opportunity. Getting to North Georgia, being a cadet here, Delta company right behind us, it took three years of kind of waiting around, and then thankfully Dr. Wells started a program, and they were looking for guinea pigs, so I raised my hand and said, yeah, I'll go to Monterey, California for two years; that sounds like a pretty good option. And then it was basically 12 hours a day of Chinese for two years, and that's a very hard language, I will say that. I came back, and thankfully another thing that they did is they signed up the first foreign exchange program with Ching Hai University, China's number one university. And I went over there because I had taken in the language; in the cultural side of the house, I was able to study with the locals doing Chinese foreign policy in China.

Guess what? They have a different view of the Korean war than we do. There're a lot of different things, so you've got to be in-country; you've got to spend time with those people to really understand. If you want to be a military intelligence professional or a national security professional, you've got to spend time with the people that you're supposed to be able to analyze and predict their decisions. The other aspect to that is, so I graduated from North Georgia here. I packed up my U-Haul, and I drove immediately to Washington, DC. I had a security clearance, I had two foreign languages, I had in-country experience, I showed up to be a translator. And guess what? I still wasn't good enough. You've got to spend time with it.

So, I had a very low-level language translator job, and I said, well, I've got to be better. You've got to spend time with it. Hard problems take time. So I packed up and moved to China, and I got a job at a local hospital in the IT department; basically, I was the only foreigner there. Because if you really want to understand how a group of people think, get as local into the group as you can; so working side by side with 32 Chinese nationals in the local hospital is a really good way to kind of figure out how they think and get that experience.

Finally after I moved back to the states at 25, I basically started cultural immersion and language. Starting at 12 years old, I'd spent more than half my life before at 25 I was finally qualified to be a translator or in the US intelligence community. Again, hard problems take time. Thirteen years just to be able to get to the point where I could effectively capture what they were saying, get what did they actually mean, what were they going to do, and that was just a base capability to start a real journey towards what, hopefully, I'll one day be an expert in.

So, again, hard problems take time.

The second thing I want to talk about is how the military doesn't really reward specialists, and pretty much that means officers. I enlisted here as a as a Chinese interrogator. In the enlisted realm, at least in the army—I'm not going to speak for other services—you can somewhat focus your career into that, but on the officer side, we are taught to be essentially managers of people who actually carry out action. There is a program called the Functional Areas Program where the Army Strategists Association, I think their functional Area 59, the foreign area officer program FA48, you can't actually apply to that until you're, I think, a major or maybe a captain, promotable but a major. Well, by that time, you're basically halfway through your career. For a thing where you can retire at 20, we've wasted half the career. Maybe it's not a waste, but if you really want a true expert, I think you need to be able to create a talent pipeline and allow for a pipeline that promotes with peers or a head of peers to create a specialist program that doesn't start when you're a major at the earliest.

I thankfully happened to work in in the intelligence community for a few years, met a wonderful person at DIA who got me into the attaché program as a lieutenant. I was the youngest military attaché in the US military for I think four or five years at one point. I have seven support tours to DAOs in east Asia southeast Asia, so again getting in-country experience working directly with the PLA or the Hong Kong police or where I'll say Singapore a few times, New Zealand, and a great tour as the assistant army attaché to Taiwan.

Because I was able to do those young—I'm only a major now, so I will caveat that now—I was able to essentially bring that back to my unit currently at army futures command to bring somebody who can actually I think effectively, at least have a good assessment on the PLA mindset and their technology transformation program. So with that I think that's pretty much my two major points. Hard problems take time, and we should reward specialists as early as possible. Thank you.

Dr. Michael Lanford:

Thank you. I found that interesting. Higher education is known for rewarding people almost over specializing, but at the same time, we have a problem rewarding cultural experts, too, I would argue. So the next speaker is Dr. Crystal Shelnutt whom I'm quite familiar with in terms of her work. I got to sit on her dissertation committee; she earned her doctorate from the University of North Georgia just this past fall, and she did a fantastic dissertation on the issues of implementation, the barriers essentially to innovative degree programs within higher education, and currently she's a senior lecturer at the University of West Georgia, so Dr. Shelnutt.

Dr. Crystal Shelnutt:

Thank you, I appreciate it. And thanks for having me, I'm learning so much; I truly am.

I am a senior lecturer at the University of West Georgia. I teach in the University College, and I teach primarily professional and technical writing to STEM students. Next slide please: here's an overview of the points that I'd like to cover today very briefly. When we talk about innovation in higher education, experts usually speak in terms of revisioning practices and policies through technology, or they speak through credentials focused on workforce development. Any innovative programs adopted should seek to produce better informed citizens, however; strengthen democracy; contribute, of course, to economic success; and ensure social justice.

To that end, and according to a 2021 AACU report, employers across all sectors, military and other sectors, private sectors, identify learning outcomes essential for success in the 21st century, and they include teamwork; digital, conventional, and quantitative literacy skills; as well as interpretive analysis. So innovative programs seeking to affect those particular learning outcomes actually have to do that by cultivating specific competencies in their students, and the first competency is intercultural awareness. Now you all have spoken today about cultural awareness, but I want to take that and make it writ small because culture does not necessarily have to extend to a different country, right? A culture can be within a particular corporation, and there are many entities within the United States that actually serve the nation's interest, as you all would agree.

So when we speak then about cultural competencies, we're talking about understanding those integrated patterns of behavior, values, rituals, communication, what have you. But when we teach our students to be aware of these competencies, these cultural literacies will equip them to work on global teams—to collaborate, to problem solve. It will also help them to navigate the new ways of thinking and the new ways of communicating because we are digital, aren't we? And of course, most importantly, those cultural competencies can help support and advance diversity, equity, and inclusion initiatives both within a corporation and on a global scale.

The second competency involves critical thinking, and, of course, critical thinking involves the imaginative process, and I'm sure that you incorporate that in in your learning modules, but I'm going to illustrate this by way of a story many of you may have read in a recent *Atlantic* article written by Elliot Ackerman. Ackerman is the co-author with Admiral Stavridis of the novel *2034*, and he was in Ukraine recently. He had met an individual—I'm going to be careful to capture this correctly in the right language—an individual who'd been fighting, and this individual shared with Ackerman his assessment of the Russians and some of the fault lines that are appearing. Now this was a couple of weeks ago before some of the atrocities that we've just seen have been amplified, and his contention was this: that some of the breakdowns are because the Russians in the battles do not have imagination. That, in other words, it's less technological variables and more psychological fit it speaks to than that to critical processes, those critical engagements, those imaginative processes. And in teaching these processes, this imagination if you will, it helps us to spot those black swans, doesn't it?

The third competency I want to talk about just briefly is communication, and I'm sure that many of your modules, your degree programs and otherwise all also speak to the idea of articulation as well as written competencies. So if those are some of the competencies that inform the outcomes that bring success in the 21st century for our students, then why are we not consistently involved with innovation? Why are we not consistently implementing innovation? Here are some of the barriers that I discovered in my research over the past three years. There was a programmatic development and implementation in the University System of Georgia, and I went through and researched across several campuses, I believe it was four or five: what are the barriers to effective design and implementation?

The first barrier that I discovered in my research was the idea of organizational change. So, if you want innovation, you have to be aware of and be sensitive to organizational change. That is, when faculty governance structures are removed that results in a lack of openness and collaboration and transparency and, as a result, leadership can become synonymous with didacticism that may in fact create schisms.

The second barrier involves this restricted view of the nature of higher education. So respondents in my study cited this rush to innovation as a focus on training versus a focus on educating, and we had talked about this last night as well. The idea of educating not just for one job but for jobs all throughout your life. So programmatic programs that are focused on this rush to innovation can be perceived primarily as enhancing students' rush to fulfill the skills gap, but then as experts will say Higher Ed becomes a center of marketing. It becomes a point where we're commodifying education, and we're rendering a product rendered by sheer economic interest.

So the third barrier that I discovered in my research concerns a little bit more of an extension of that discussion and that is the transforming power of education. The respondents in my study resisted the idea that we would in Higher Ed, in any realm, focus on mere scaling of efficiencies because when we do that, we neglect that transformative power of education. Often as educators, we don't reap the benefit of our involvement and investment in students for many years to come. It's not an either-or proposition here, and it's not a one-to-one ratio. When we're developing good citizens, when we're strengthening democracy, when we are in fact concerned with economic interests, those simply may not appear immediately.

In fact, Admiral Stavridis—once again, just a fascinating guy who loves liberal arts may I say and reads 100 books a year—he contends that we can combat this tendency in Higher Ed if we will recognize that education in the military is both a value and it's valued. That it's not a one and done scenario in the military. That Higher Ed actually can take a page out of the military's book by continuing to look at education as that continuous cycle of learning as programs that demand self-responsibility in learning and that enact structures to aid those career trajectories as opposed to just the one implementation process.

Why does all this matter? Why do we even care? To amass the portfolio of skills that I described demands innovative programs at your university, at mine, and all the way through Higher Ed.

However, again from my research, I discovered that to usher in innovation takes work, and it takes time; it does. It has to be inclusive. That is, faculty have to be involved.

Certainly, we need SMEs; we do. We absolutely have to have them conjoined, but faculty have to be involved. Programmatic development needs to adhere to the distinctives of higher education.

We hear much today about faculty governance and the dismantling of tenure, but those structures have been in place for a very long time and have done this in very good stead for that long time. Programmatic development should also develop that personalized change process, and that change process should include looking for institutional entrepreneurs and relevant change agents within the institution. They are there, and they need to be brought on board and be part of that structure. Programmatic development also needs to focus on creativity and balance. We need to reward and incentivize innovation, and we have not always done that in Higher Education.

The second conclusion here is that developing innovative programs requires partnerships, and I just loved listening to General Brown today. He really kind of stole my thunder because what he's talking about is developing these kinds of partnerships that are mutually beneficial and they're mutually respectful. We've lived long in a milieu where we are at odds with one another, and it need not be so. So one study on intelligence and programs in Higher Ed and beyond, in 2010, says that this liberal education, these competencies that I've discussed, that I've mentioned to you, actually provide the soundest footing and foundation for higher intelligence programs. And in this study, this author calls for closer collaboration with Higher Ed to continue to develop those. In fact, Jamie Merisotis from the Lumina Foundation, talks about the need for this virtuous cycle of education where can work and national security can collaborate, and they can help create paths to help a lot more people in which everyone can learn, earn, and serve.

So his point of contention is that we can do all of these things with the aim of empowering the economy. We can do it by strengthening our democracy and advancing equity and justice. And then ultimately innovation requires complementarianism. So all sectors, and particularly the military, need humans. You need humans to use their own competencies and their own abilities and capabilities with AI and other high-tech advances. It's not an either-or proposition. We have to have individuals who employ their ethics, their compassion, their passion, their reasoning skills, to provide cohesive structures and substructures of reason, evidence, and assumptions. Therefore, Higher Ed must yoke the learning trajectories to these emerging technological systems. Once again, we need to develop individuals able to detect those black swans. So as former congressperson David Skaggs maintains, national security is inherently a function of the economy, and the economy is inherently a function of educational attainment. Thanks for listening.

[See Appendix for corresponding PowerPoint presentations.]

Dr. Michael Lanford:

Thank you, Dr. Shelnutt. Very comprehensive. We have a lot to talk about with your topic today, thank you.

Iyonka Strawn-Valcy has had an extensive career in promoting internationalization. She's worked as Director of Global Operations at Georgia Tech most recently, and she is primarily involved with Georgia Tech's campuses in France and Shenzhen, China. She's also, I'm proud to say, a student in our UNG doctoral program where she's planning to do a policy analysis along with interviews of senior internationalization officers at research intensive universities. So I'll yield the floor to you, Iyonka.

Iyonka Strawn-Valcy:

Thank you so much. I will start by saying that, in terms of stealing thunder, I'm going to dovetail off of what some of my colleagues have already mentioned. Hard problems take time... that that was great. Thank you for sharing and reinforcing that but also teamwork and cultural competencies as well as innovation—and that is the focus of what I want to talk about today, specifically related to higher education and comprehensive internationalization.

The perspective I'm sharing on this panel comes from an institutional internationalization standpoint, and I'll draw from this quote here that is part of a joint statement of principles in support of international education between the U.S. Department of State and the U.S. Department of Education. In terms of tying all this together, what I want to focus on is that the U.S. cannot be absent from the world stage when it comes to these sorts of matters. When it comes to world challenges, in terms of improving the human condition, which is what we talk a lot about at my institution, working together—partnerships, collaboration—is a hallmark of what campus internationalization is, and you'll hear me talk a lot about committees, working groups, task force, and lots of other groups that we have that we to try to bring together different partnerships and collaborations related to leadership and internationalization.

So internationalization has largely been impacted by globalization, and I won't get into the various definitions and theories related to globalization or we will be here all night, but I just want to highlight that in terms of some of the strategies that have come up with as a result of internationalization which I'll talk about today. And we have study abroad, which I'm sure a lot of you are familiar with, but also international research, collaborations, international student and scholars studying here in the U.S., and the internationalization of classrooms, so internationalizing the curriculum as well as international branch campuses which is what I work with quite a bit at my institution. And in terms of partnerships and collaboration, I'll give an example: during the pandemic when we had students who were unable to travel for the most part based on a lot of different scenarios and concerns that we all dealt with, but in terms of looking at our partnerships, our campuses and in France and in China, we were somehow able to try and tie together as many of our collaborations and resources as possible to ensure that our students stayed enrolled so that they didn't have to withdraw for a year or more, but using our international partnerships to do that. You don't have to have a gigantic enterprise of branch campuses and off-site campus locations and educational facilities to be able to do that. You can have reciprocal partnerships with international institutions; you can have partnerships also with industry as well as non-governmental organizations—but focusing on that and leveraging those partnerships to support access to also underpin U.S. leadership.

In terms of those partnerships, something else I wanted to mention is increased cooperation between the federal government, the private sector, and educational institutions to maintain the integrity of federally-funded programs and also to protect intellectual property and research endeavors from undue foreign influence and unlawful acquisition. Part of the strategy that we work with at my institution is to understand that we do need subject matter experts as my colleague had mentioned; we do need faculty to also weigh in on these situations; and we have many cross-cultural or cross-functional teams that help to support a lot of the work we do. We cannot do it alone, and pulling from those different resources is really helpful. We have an international crisis working group which is something I wanted to bring up because it's something that's new that started developing, I would say, during the pandemic.

So a lot of times when we have situations where we have students traveling abroad, there may be various groups to review and to approve or to vet travel, but we found what we did not have is sort of a body to look at international crises and say, well, how can we be responsible citizens and make sure that we're addressing the global needs but also the local needs that we have for our campus community and for our local communities as well? And so that's how this committee and this working group kind of developed, but what we've been able to do is also leverage these conversations in in other situations where we were trying to address situations with refugees from Afghanistan, for example. And right now, a huge focus of this international working group is how do we support our students from Ukraine, how do we support our students from neighboring countries as well that may be impacted by the crisis going on there right now? How can we work with the local government? How can we work with other organizations to help support these needs? And pulling from various, again, individuals who can help provide some sort of support, even raising funds to ensure that students can stay enrolled.

So these different working groups have lots of cross-functional responsibilities, and the last I'll say about that and what I want to emphasize on this slide is that strengthening relationships between current and future leaders to provide national and global leadership is a huge benefit of higher education internationalization, but it also impacts how the U.S. is perceived globally, which we feel is really important. So this international crisis working group is stretching out hands to as many different organizations as we can to support our students and our faculty but also looking at who else can we support in this endeavor, if we have the resources indeed to do so, who else can we support?

In terms of threats to internationalization, I know that we've probably discussed some of these things in the various talks that you've heard today or perhaps tomorrow, but I just wanted to highlight what some of these are and use an example of another group that I work with at my institution that helps to advise our executive leadership on strategy in terms of some of the geopolitical events or economic challenges that may be occurring in the U.S. at this time and also globally. What we try to do is be risk averse, also understanding that sometimes risk averse means shutting ourselves off from everyone else, and we have so much to do at my institution and all institutions related to research, teaching, and learning that we want to make sure that we share that where we can and not shut ourselves off completely from the rest of the world. So what we like to say is that we are risk aware and not necessarily risk averse, which can help us to understand what the challenges are and bring people into the conversations that will help provide strategies but also say, how far can we address the situation? Or how far can we extend our internationalization goals with being in compliance and making sure that we are in compliance and that we are exercising safety and risk management as well?

In terms of challenges and threats to internationalization, emergency program funding is probably one of the highest areas of priority that I tend to work with but also, as I've mentioned, committees and work groups to address operations, risk management, and international crises. Virtual learning, as we've seen, has become extremely important in recent years especially, but it's been around for a very long time in fact, and it's been heightened as a result of the pandemic, but it's definitely something that's worth leveraging when it comes to internationalization. And then, of course, innovation, personnel, and hiring. At my former institution, we developed a position for an international safety and security officer, and we were very lucky in that moment to have been able to hire someone who was a former secret service officer, and this person came in to help us strategize but also analyze the various positions that we were in—situations we were putting our students in—locations where we had programs to really help advise on travel... what was something that we could possibly do or something that we maybe needed some additional support and funding for? And a big piece of my job is to make sure that we get the buy-in from our executive leadership to know the answers to these questions: what do we need funding for? what do we need support for? If internationalization is going to be a priority, how can we continue to leverage that?

In terms of innovation, research, and teaching, I wanted to make sure that we mentioned this in terms of the growing importance of university partnerships networks and collaborative initiatives with international organizations. The development of branch campuses and an expansion of agreements and agreement dynamics is incredibly important, but it's also incredibly complicated. So when you have institutions in a different cultural setting, when you have an educational facility in a different cultural setting, how do you navigate that—where you are fulfilling the purpose of your educational facility but also taking into account the cultural values and the ecosystem within the country or the culture where your institution is located? So that's something that's always a constant area of conversation and just critical analysis in my office... in terms of making sure that we leverage what we can leverage but also being good partners and being good stewards of the funding that we have to develop these branch campuses and make sure that we're doing it ethically and sustainably.

Finally, I just wanted to share some strategies to threats in international research collaborations. Undue foreign influence is something that's becoming more and more of a topic when you talk about international research, and making sure that you have strategies to work with that is really important. Recognizing the potential conflicts of interest and conflicts of commitment which, of course, are two different things. Insisting on transparency, reciprocity, and adherence to research integrity when engaging in collaborations or when hosting scholars and complying with rigorous disclosure affiliations and commitments, and my office has a very strong collaboration with our legal team and our office of ethics and compliance to make sure that we're consistently offering programming to our students, our faculty, and staff related to, what is undue foreign influence, first of all? And what does that mean? And, then, what is conflict of commitment? How do I know when there's a situation that may involve some sort of a conflict of commitment, and what are the resources that I have in order to be able to manage that and to navigate through that?

I just want to recap that employing a coordinated national approach to international education is one of the best ways to support higher education internationalization and national security,, and a lot of time, it starts at the national level, then the institutional level, then the program level, and then the interpersonal and interactional level. That's all I have, thank you.

[See Appendix for corresponding PowerPoint presentations.]

Dr. Michael Lanford:

Those are wonderful insights from a part of higher education that we don't often hear as much about but is becoming extremely important over time especially, not just even at research universities but across the board to regionals like UNG and even community college space increasingly, so thank you.

Our fourth speaker, who is meeting with us virtually, is Dr. Magdalena Bogacz. She's an Assistant Professor of Leadership and Ethics at our university, and I got to know Dr. Bogacz through a mutual friend. She shared some of her research, which was just fascinating. It basically concerned the barriers to greater gender equity in higher education and why we don't have equity among faculty, and Dr. Bogacz's research in particular focuses on how Ph.D. programs often discriminate against women, so I'll hand this over to Dr. Bogacz and thank her for Zooming in.

Dr. Magdalena Bogacz:

[Please see peer-reviewed article entitled "Linking Innovation Back to National Security via Innovation Ecosystems: The Role of Higher Education and Equitable Faculty Socialization" in this collection.]

Yes indeed thank you Dr. Lanford for this gracious introduction. I'm very excited to be with you today virtually. Unfortunately like some other members of our symposium, I wasn't able to make it to Georgia yesterday due to severe weather conditions in Montgomery, Alabama so we'll have to do that on Zoom, and before I dive deep into my slides on the topic that is very close to my heart, as Dr. Lanford mentioned, I would like to very quickly share a disclosure that opinions, conclusions, and recommendations expressed in my presentation are mine and mine alone and they do not necessarily represent the views of the our university or the United States Air Force. And with that in mind I would like to ask for the next slide and let's get to it!

So the first question that you might have or wonder about of course is… what exactly is faculty socialization? And to put it very simply and briefly, it's a process of how faculty learns to be faculty, and it consists of processes, practices, and policies in educational organizations that allow to transition faculty from being organizational outsiders to organizational insiders. And there are two stages to that process: there is anticipatory stage that usually occurs prior to employment during graduate school and applies to graduate students, and then there is a second stage, organizational stage, that occurs after employment is completed during early onboarding and new faculty mentoring programs. And this specifically applies to junior faculty members… so fresh graduate students that are becoming faculty members. So now knowing what faculty socialization is, we may further ask if there are some kind of forces or factors that can potentially influence faculty socialization and affect it in a positive or negative manner. And of course the answer is yes because faculty socialization processes occur within complex systems such as educational organizations, they will be inevitably affected by a given organization's context and culture—all the good the bad and the ugly.

The situation is similar with faculty socialization processes that faces challenges that disproportionately affect underrepresented populations such as women, people of color, people differently able, and international students, which I think is particularly important for professional military education and military in general. And there are variations within academic fields, and what I mean by this is that some fields experience greater challenges with faculty socialization processes than others, and that should come as no surprise that stem fields are some of those disciplines that face greater challenges with how they prepare graduate students and later junior faculty members to be the best academics that they can possibly be, but there are also other fields that I think often go unnoticed, and those fields are humanities such as philosophy, where we teach critical thinking. Today we heard a lot about critical thinking . . . how can we equip students to be better thinkers, more adaptive, more creative, more innovative? We often do that in philosophy. Another field where some groups of students and faculty members are disproportionately affected by faculty socialization processes is theology and logic, or even ethics or history. And it's not just those populations are underrepresented in those fields. That's one big problem, an issue that we're facing, but those groups are also underfunded, underrated, under-rewarded, and later under-promoted, once they become faculty members.

So why is this problem significant, right? Why is talking, discussing, and trying to improve faculty socialization processes significant? I listed five considerations that I think are our most important, and I don't think I have time to dive into all of them in greater detail, but I will focus on two in particular the last two: quality of higher education and enhancement of our national security. I would like to argue that by increasing or making our faculty socialization processes better, more equitable, more just more fair, more inclusive, we are improving the quality of higher education, and a specific aspect of quality of higher education that

I as a philosopher am most interested in is cognitive diversity. And I would like to add to this that I believe the collective intelligence of academia comes from heterogeneity of its members: from free exchange of ideas, from being open to disagreements, from a variety of viewpoints, from multi-disciplinary perspectives. By improving our faculty socialization processes and uplifting those critical underrepresented and often under rewarded and unheard voices, we are simply making our education better. And why should we care about making our education more diverse and better of higher quality? That's because there's this strong link between higher education and our national security. Academia that excludes groups of individuals is limited in scope at best, but it's unreliable at worst because it produces limited knowledge. So what this means is that our security environments that academia inevitably takes part in and helps to produce those comprehensive global interconnected, full security environments, is lacking innovation creativity and diversity in order to produce objective knowledge and provide us with those objective full comprehensive security environments. So, this ultimately impacts our national security.

That is why it's important to think about diversity as a strategic advantage and there's no doubt that our adversaries do that, and so the link between faculty socialization, diversity, and national security is innovation, the topic of our panel, there is a wealth of research on innovation that shows that for organizations to become more innovative, our academia to become more innovative, we need to be open to new ideas we need to be open for multi-disciplinary perspectives, we need to be welcoming for those critical underrepresented voices that often go unnoticed and forgotten where the talent and gifted individuals slip through the cracks. Innovation ultimately requires diverse backgrounds, and although we do have the best educational system in the world, I believe there is still a lot that we can do, and we can certainly improve innovation in our higher education institutions.

Justification for focusing specifically on faculty socialization processes for me comes directly from the joint chiefs of staff, and in the 2020 document entitled "Developing Today's Joint Officers for Tomorrow's Ways of War", the joint chiefs of staff presented their new vision and guidance for not only professional military education but also higher education in general, which we already know they are very much interconnected, and they mentioned that their new vision for fully aligned PME and talent management system relies on identifying, developing, and utilizing strategically minded, critically thinking, and creative joint warfighters. So what does this mean for higher education and professional military education? It means that we have to promote and rely more on innovation creativity original thought and cutting-edge research all to compete on a globalized interconnected arena to face the return of the great power competition and to face the constantly changing character of war.

When it comes to actionable deliverables all solutions because this research is in its early stages I do not have specific recommendations of what exactly can we do to improve our faculty socialization practices and policies in order to increase our cognitive diversity and therefore enhance our national security, but I have some general ideas that I think could be easily adopted by a variety of different institutions to fit their needs, but I think one aspect of those solutions that is important to focus for us today since we are working on making those connections between higher education and national security and military apparent, would be is there is anything in particular that military can do to support higher education innovation? And I believe there is. Unfortunately and fortunately, I will have to echo many of the speakers that came before me including General Brown that the idea is we need to work on creating supporting and funding initiatives that encourage collaborations and partnerships not just between private and public sectors, but also between civilian and military sectors as well as our national institutions and international education institutions of our allies. In other words, there is power in international relations multicultural competencies.

That will be a conclusion that I am going to leave you with today, and that is what are the implications of improving our institutional processes with how we train, prepare, uplift, and mentor faculty on both levels graduate and early employment. And the implications are that by improving these processes and making sure that they are more inclusive, more creative, and innovative, we would be ultimately increasing cognitive diversity in our educational institutions. I think that's important because by this we will create a more objective and comprehensive learning and teaching environment which will generate a more comprehensive knowledge about security environments, which is particularly important in today's globalized world. Also, by including more diverse voices among our educators both in higher education and professional military education, we would be able to better gather, analyze, assess, evaluate, and disseminate information in a more inclusive, global, and complete fashion. And I think that ultimately this way of producing sharing knowledge would better align with the vision and guidance of education as presented by the joint chiefs in the 2020 document. That will conclude my presentation today, so thank you very much and I hope to connect during the q & a after the panel finishes thank you.

[See Appendix for corresponding PowerPoint presentations.]

Dr. Michael Lanford:

Thank you so much, Dr. Bogacz. And for our final speaker, I want to thank our actually two speakers, but we're going to have one of them up at the panel here today, for coming all the way from Poland. Professor Jacek Dworzecki from the University of Land Forces in Poland will be speaking about links between the military and Poland and higher education.

Professor Jacek Dworzecki:

Thank you very much. As Dr. Lanford mentioned, I represent the Military University of Land Forces in Poland. Sorry for my English, but I'm living in Poland, and I'm teaching in Slovak, so I speak in three languages, but, unfortunately, I don't speak fluent English. But maybe in the near future I will speak better.

In the frame of my short presentation, I will try to show you our point of view. However, that means if I say "our," that means Polish people, Slovak people, and Czech Republic people. This is a population of approximately 55 million people, because in Poland live 37 million people, in the Czech Republic, 10 million people, and in Slovakia, 5 million people. So, as I said at the beginning, I'm working and teaching in these three countries many years. The situation that I'm addressing is this war, the invasion of Ukraine by Russia. It's extremely hard for our population. I don't speak about the point of view of Hungarian people, because as Hungary takes a different point of view from us because, we must say this, they support Putin in his war. But another thing is, we can't say Putin's war because 80% of Russian society supports their president in this aggression, and for us, it's an unbelievable situation. I'm 46, so I was born in and grew up in the Communist era, with the communist transformation in Poland which occurred was in 1990, so I remember this time. This was a wonderful time, from point of view of Russian people, was when my father waited four days to buy petrol, when we waited in long lines to buy a sugar or chocolate. And now we hear that the Russian Federation wanted to liberate us from the Nazis. They found Nazis in Ukraine, in Bucha and in other cities where there were snipers killing five year old children because they wanted to break the civilian Ukrainian resistance. This is our situation now; it's an extremely hard situation.

This symposilum is being held at a military college where you teach U.S. soldiers. We, too, knowhow duty is for soldiers. The soldiers must protect the government, must protect the country, must protect the nation, but how do we explain today's current situation where the Russian, I can't say soldiers, so will say these people are attacking civilians; they're robbing, they're raping, and they making genocides. The effects for our part of Europe, and from European West, European Union East countries, that means Poland, of course, Lithuania, Latvia, Estonia, Poland, Slovakia; we're really afraid because, yes, of course, we have extremely strong support from the USA. The USA is our biggest supporter as well as, of course, NATO. But how is its possible that one man from the Kremlin can make these kinds of choices and attack sovereign, independent countries?

As you know, Ukraine is a big country like Poland; its population is 40 million, with 40,000 square kilometers, and Putin wants take hold of this country in 2022. So, in Poland, we are now changing direction to protect ourselves, and our government wants to increase our budget for military purposes that next year will be 3% GDP— that will be approximately 20 billion dollars. It's not enough; it's not too much. We try to make our army bigger. Now in Poland we have only 125,000 soldiers, so the main goal of the Polish government over the next three-to-five years is to change our structure, our army, and prepare the professional army for 300,000 soldiers. Of course, we need support; we need equipment. We try buying super equipment from you. We spoke about tanks and fighter jets, but also we need strong support from European Union countries, from main players, from Germany and France, and we don't see this support because the mode of business as usual is still going on in West European Union countries. We in Poland, are not hardly addicted to Russian resources, to Russia LPG gas and oil. But in, for example, Germany or France, the situation is different. And Hungary is buying approximately 91 or 95% of their gas from Russia, so they are extremely addicted, and they try to block every sanction now.

In our paper, we wrote about the participation of Polish Military Universities in education for security, as a task of the Polish national security system, so I can talk about this in this moment. Maybe I can say about the nearest challenges for Polish society, for Slovak society is our common border with Ukraine. It's approximately more than 650 kilometers because we have 560 and Slovakia have only 90 kilometers, and we have also a common border with Belarus, which is a right hand of Putin's, and, of course, we have a common border with the Russian Kaliningrad district on the north side of Poland.

So now we are thankful for the USA, the American nation for supporting us because we sent more soldiers to Poland; it's probably now 10,000 soldiers with super modern equipment. But we must prepare our army. And the second thing, as you see in the Ukrainian-Russia war, the second important thing is direct support from civilian guards or police support for the soldiers. Ukrainian police is very effective and flexible in the frame of fighting with Russian sabotages, and we want also to create a new reality for Polish police officers. We have in Poland 100,000 police, and we will be training them for use of military equipment. Of course, that will be possible by using model simulators. For example, how to use javelin or NLAW, how driving a striker, you have russell mac, but you have striker heavy equipment, so that will be the challenge.

And, of course, we educate Polish soldiers, and we have four military academies in Poland. Our academy is mainly for infantry and tanks, so we try to create for them the best opportunities and possibilities to learn and grow up in their professional duties. We have now in Poland only professional soldiers, so when they come to our university, they are as a soldier in uniform, and they have a monthly salary, so they have good opportunities to grow up in their duties. That means approximately 15% of Polish soldiers can be officers. In the police, this level is lower; approximately 25% of the police can be officers. That means better salary, better opportunities, better protection, and the nearest challenges is also to help the Ukrainian nation to stop this war. I don't know if it is possible to win this war, but we try help them with families.

We have now more than four million Ukrainians in our country; this is 2.5 million refugees. And free and after attacking Crimea in 2014, many Ukrainians from this side of Ukrainian, from east side, they came to us. So now, 10% of the Polish population are Ukrainian refugees, mostly seniors, women with children. So we don't create from today any kind of camp for refugees. Polish citizens take these refugees to our own homes. This is only a month, but the war will be long on the east side of Ukraine; the war will be at a minimum a year, say, in general, a minimum of a year because the Ukrainian nation—it's not a nation where people accept comrades from Russia. They will be fighting with them. Ukrainian people want to be a part of Europe; they want to have a democratic government; they want to have better opportunities; they want to have a better life. And Russia liberators, they won't win this war, but probably they will destroy the east part of Ukraine, Donetsk, Luhansk, Mariupol. Thank you.

Dr. Michael Lanford:

I just learned a couple weeks ago that my thesis advisor for my master's degree is from Ukraine. He just learned at that time that his son who is at the University of Toronto flew back to Ukraine to help fight, and we see this all over higher education right now, and institutions throughout the world are being impacted. We need to think about migrants. We need to think about how to help people in a humanitarian way, and I appreciate you giving us your perspective especially in this very difficult time coming all the way over here. It's a lot of issues to deal with.

4B

Linking Innovation Back to National Security via Innovation Ecosystems: The Role of Higher Education and Equitable Faculty Socialization

Magdalena T. Bogacz
Assistant Professor of Leadership and Ethics at Air University's Global College of Professional Military Education

[underwent separate external peer review]

Abstract

In the past three decades, there has been a shift in the US innovation environment from public to private-sector-led innovation. The change has been largely driven by the expansion of the commercial market for technology. Consequently, companies and organizations started to focus on commercial innovation, rather than on national security, and the Department of Defense became an innovation consumer, rather than a producer. Within the national innovation ecosystem, however, one establishment could quickly link innovation back to national security: higher education. The nation's colleges and universities, with their abundance of talented researchers, could effectively drive innovation and entrepreneurialism focused on national security. This paper argues, however, that innovation for national security requires the development and preparation of diverse faculty, as well as the cultivation of meaningful and trusting relationships with the government. One way to do so is through equitable faculty socialization that embraces the distinct advantages inherent in multicultural society.

*Keyword*s: faculty socialization, higher education, innovation, innovation ecosystem, national innovation network, national security, talent management

Author Note

Magdalena T. Bogacz is an Assistant Professor of Leadership and Ethics at Air University's Global College of Professional Military Education.

Opinions, conclusions, and recommendations expressed or implied within are solely those of the author and do not necessarily represent the views of the Air University, the United States Air Force, the Department of Defense, or any other US government agency.

Correspondence concerning this article should be addressed to Magdalena T. Bogacz:
Email: magdalena.bogacz@au.af.edu
Phone number: 661-496-0755

Introduction

Multiculturalism has become clamant in most United States academies after 1970. The idea that historically disadvantaged populations, such as women and racial, ethnic, and religious minoritized populations, have different and valuable insights and ways of producing knowledge gradually permeated academic circles, course curricula, and research (Asante, 1996; Mitchell, Nicholas,

& Boyle, 2009; Song, 2020). The revisions of curricula that took place from the elementary to the university levels were designed in hopes of correcting what was perceived to be an incorrect and unfair Eurocentric perspective that put too much emphasis on the contributions of colonial powers while underemphasizing the contributions of indigenous people and people of color (Eagan, 2022). As the academe became more diverse, the resulting discourse quickly augmented scholarship on innovation, as well as the ethics of teaching and learning. This newly formed outlook about the importance of incorporating different cultures, languages, and diverse perspectives in all domains of education led to the creation of a number of specialized areas of study and departments that concentrate on civilizations, traditions, and heritage from specific geographical locations (Banks, 1995; 1997; Education Encyclopedia, 2022). In order to acknowledge the distinct advantages of multiculturalism, that is, to continuously create and legitimize knowledge from different lenses and perspectives and to challenge prevailing stereotypes, colleges have made strategic efforts to diversify their student and faculty populations. One way to successfully ensure that individuals from minoritized groups feel supported by elite institutions with customs and traditions grounded in a white, Judeo-Christian heritage is through a process of faculty socialization. Hence, this paper suggests that effective, efficient, but—most importantly— equitable socialization practices can help with hiring and retention of women and minority professors, thereby embracing the ideology of multiculturalism. This is important because empirical research has shown that diverse faculty produce innovative theories and scholarship through diverse social interactions, and such an environment would also breed scientific and technological research necessary for national security (Lanford & Tierney, 2022). Hence, this multicultural environment is necessary to maintain and increase our nation's competitive edge in today's ever-changing academic, economic, political, and military landscape.

THE SIGNIFICANCE OF FACULTY SOCIALIZATION

The term *faculty socialization* refers to the process of "how faculty learn to be faculty" (Tierney & Rhoads, 1993, p. 5). More broadly, faculty socialization consists of processes, policies, and mechanisms that help faculty transition from being an organization's outsiders to insiders. Under ideal conditions of socialization, the organization also changes some of its structures and processes (National Research Council, 1997). Tierney and Rhoads (1993) further suggest that the processes of socialization can be divided in two main stages: an anticipatory stage and an organizational stage. According to the authors, the anticipatory stage occurs prior to employment, usually during graduate school when students are first exposed to the norms of the professorate (Tierney & Rhoads, 1993). Students then learn the basics of faculty life and contextualize this new knowledge through their past experiences. The organizational stage occurs after a doctoral graduate obtains faculty employment, during onboarding and new faculty mentoring programs. This stage is when junior faculty learn how to adjust to, adopt, and—in some cases—challenge institutional values, beliefs, and norms. Because the early years of faculty life are challenging, stressful, and alienating, many new faculty leave academia within the first two years (Boice, 1991, 1992; Crepeau, Thibodaux & Parham, 1999; Korte, 2007). Moreover, inadequate, or ineffective socialization can exacerbate faculty exodus. This is because, as Korte (2007) points out, socialization affects employee's attitudes and behaviors, and it is a routine process for an organization to share and maintain its culture. In other words, it is through effective socialization practices that junior faculty retention, performance, and job satisfaction can be increased.

Additionally, extant research suggests that faculty socialization processes have a disproportionately negative affect on underrepresented populations, such as women and people of color (Kelly, McCann, & Porter, 2018; Johnson & Lucero, 2003).

Moreover, professors in some academic disciplines experience more socialization problems than others. For instance, science, technology, engineering, and mathematics (STEM) disciplines, as well as philosophy, logic, and ethics fields, are generally more prone to alienate historically marginalized populations (Bogacz, 2021; Posselt, 2020). Within these subjects, minority groups are underrepresented, underfunded, underrated, and under-rewarded (Hutchison & Jenkins 2013; Kahn & Ginther, 2017). Furthermore, faculty from minoritized groups frequently have limited access to professional mentoring programs and support groups, while experiencing disparities in access to post-graduation placements (Thomas, Willis & Davis, 2007; Zambrana et al. 2015).

This is a tragic state of affairs, as scholars from historically underrepresented populations have an abundance of talents, unique perspectives, and valuable experiences that can contribute to the production of original thought and, therefore, increase innovation and cutting-edge research. As a result, higher education policies and practices must be more responsive to the problems of underrepresented minorities. The way the faculty advisors teach, train, and prepare graduate students to join the professoriate matters no less than incorporating junior faculty into their newly adopted organizational cultures (Bauer et al., 2007). More equitable training of students and junior faculty requires a shift in mindsets and the reallocation of organizational resources to better help those who need such resources the most. This would also mean cultivating academic, disciplinary, and organizational cultures that care about cognitive diversity and fair organizational practices.

FACULTY SOCIALIZATION: A COMPLEX PROCESS

The complexity of faculty socialization comes from an interplay and codependence of several distinct cultural factors. In 1987, Burton Clark, a widely respected professor of higher education at the University of California at Los Angeles, famously described five cultural forces that shape faculty life. They include 1) the national culture, 2) the culture of the profession, 3) the disciplinary culture, 4) the institutional culture, and 5) individual cultural differences. Moreover, understanding socialization through faculty culture reveals parallels to anthropologist Clifford Geertz's (1973) "web of culture" view where culture shapes and is shaped by social interactions, events, and networks. Geertz famously said that "man is an animal suspended in webs of significance he himself has spun, [and] I take culture to be those webs" (The Interpretation of Culture, 1973). On this view, culture is a complex web of interconnected people, parts, and processes and our institutions and practices reflect certain intertwined cultural norms and values that make it difficult to change one strand of the web without affecting the entire system. Influenced by the writings of Geertz, Tierney and Rhoads (1993) inferred that socialization in higher education is an *ongoing* and *bidirectional* process. First, socialization is ongoing because even experienced senior faculty face institutional challenges and obstacles that they must face and overcome. This requires continuous learning and relearning of faculty roles and responsibilities. Second, socialization is bidirectional because faculty adapts to organizations, and organizations adapt to their members.

Tierney and Rhoads' (1993) view that socialization's being ongoing and bidirectional sets conditions for the cultivation of diverse academic contexts and communities. The authors wrote, "While professors change to meet the demands of their academic institutions, colleges and universities must modify their structures to meet the needs of their diverse members" (p. 6). This means that organizational processes—such as academic rank promotion, faculty development, and performance assessment—must be continually reviewed and adjusted to suit the needs of the evolving body of professors and align with the values, goals, and objectives of the organization. However, a perpetual question concerns whose values and goals become upheld and promoted within academic organizations and the process of faculty socialization. Majorities? Minorities? A mix of both? It is possible that the answer is threefold. First, underrepresented populations have fewer opportunities to socialize quickly and effectively, due to the complexities of cultural forces. Second, underrepresented populations' impact on organizations is less severe and slower. Third, institutional inclusion of and adaptation to underrepresented populations is passive at best, and imperceptive at worst. This posits at least two further challenges to the production of meaningful and innovative research: inequitable organizational practices and ineffective talent management.

Equitable and Culture-Sensitive Faculty Socialization

To transform the workplace and landscape of academe—that is, to make it more inclusive, promote a greater interaction of diverse cognitive talents, and create a multicultural environment that generates a comprehensive and exhaustive teaching and learning experiences—the socialization of underrepresented minorities into faculty life must become a more equitable process. The effectiveness and ease with which the socialization of faculty takes place depends largely on organizational culture. But many of today's organizational cultures were formed on historical and social patterns that excluded or otherwise disregarded the potential contributions of individuals from minoritized groups in their formative years (Barber et al., 2020; Leuschner, 2015). This, by design, puts underrepresented populations at disadvantage. If organizational cultures were initially formed without incorporating diverse perspectives and points of view, they often became mainstream, by default, and less open to the inclusion of multicultural perspectives and more resistant to the incorporation of diverse faculty into their fabric (Beebee, 2013). This is because of the very nature of bidirectional socialization, where people adapt to organizations—and organizations, at the same time, adapt to people (Tierney & Rhoads, 1993).

The evidence of this is empirical, as women and people of color are underrepresented in US higher education. This becomes clear when we examine the number and seniority of full-time faculty positions in American universities (Johnson & Lucero, 2003; National Science Foundation, 2019; Stewart & Valian, 2018). For instance, among full-time professors, 28 percent are White females, 4 percent are Asian/Pacific Islanders females, and 2 percent each are Black and Hispanic females (National Center for Education Statistics, 2022). This trend continues down the academic rank ladder with the most White females working as instructors, 42 percent, and lecturers, 44 percent, but remains steady with women of color, at between 2-5 percent at instructor and lecturer level. However, it is important to point out that unequal representation of women and racial, ethnic, and religious minorities cannot be fully accounted for or explained by just one inequitable institutional process, such as socialization. Other processes (such as hiring and retention) and other systemic factors (such as pipeline issues) play an important role in understanding what has caused the US academic landscape to look the way it does (Johnson & Lucero, 2003). It is worth noting that many colleges and universities have redoubled their efforts to diversify academic departments through institutional reforms of hiring and retention practices, but less attention has been paid to the process of socialization itself (Liera, 2019; Posselt, 2016; Slay, Reyes & Posselt, 2019; Villarreal & Liera, 2019). Since faculty socialization is an ongoing process that starts in graduate school and extends to the workplace, the faculty advisors may be able to achieve two aims at once by 1) teaching women and minoritized individuals about the basic functions and responsibilities of being a faculty member and preparing them to quickly and effectively adapt to their new workplace, and 2) improving institutional and disciplinary work environments so that underrepresented faculty feel welcomed, supported, and invested in their organizations and fields of study. In other words, by addressing the inequities with faculty socialization, we would impact women and minorities representation at the highest faculty levels, and potentially upgrade or at least enhance a few organizational processes, such as hiring and retention, that would mitigate long-term retention issues.

In addition to changing academic workplaces and landscapes, more equitable and culture-sensitive faculty socialization would potentially increase effective talent management within educational organizations. For decades now, the United States has been considered to have the world's best system of higher education. Current data supports this claim as roughly half of the top 50 institutions in most ranking systems are US universities (US News & World Report, 2022). The US especially dominates in science, technology, engineering, and mathematics (STEM) related fields, as compared to other regions of the world, including Global North countries such as the United Kingdom or Germany (Shanghai Ranking, 2021). While US supremacy still reigns true in these important fields, there is compelling data to suggest that the gap between the US and its global competitors is quickly shrinking (Independent Task Force Report No. 77, 2019). Data indicate that the US is losing its advantage in STEM-related fields where women and minorities are significantly lagging in numbers behind their white-male counterparts (Blau & Kahn, 2016; Cheryan, et al., 2017; Kahn & Ginther, 2017; Kelly & Grant, 2012; National Science Foundation, 2014a; Posselt, 2016). This is true for both students and faculty. Diversifying such a demographically—and potentially impoverished academic landscape—may mitigate the current decline in our nation's competitive edge. Hence, for the purpose of maintaining US dominance, as it pertains to scientific and technological innovation, we ought to encourage greater diversity in these academic fields which are so vital to national security. If we are to fully embrace the views listed above, then we need to focus on enlisting more minorities and women into the ranks of the new academic generation. It is in the country's best interest to increase the sheer number of talented minds who contribute to the advancement of our country, and this entails the recruitment and retainment of those who have been ignored in the past. New faculty are on the academic front lines shaping and determining how well the higher education system adapts to new realities, but it is diverse faculty that create and legitimize knowledge from many different perspectives and challenge prevailing stereotypes. Thus, ensuring that our new faculty is more diverse through improved socialization practices would give our system of higher education a greater competitive edge, a greater adaptability to constant and unpredictable change, and it would also produce greater innovation.

HIGHER EDUCATION, INNOVATION, AND NATIONAL SECURITY

From World War II through the 1990s, Department of Defense (DOD) played an important role in supporting innovative research by public institutions, such as universities. The DOD acted both as an investor and first adopter of projects that it funded and fulfilled by contracting people and companies and directing research as desired and necessary. This *government-centric* model allowed the U.S to dominate the world in innovation, research, and technology development, and the US quickly became the most secure and prosperous nation on earth (Lewis, 2021; Mazzucato, 2013). In addition, being a world leader of innovation allowed the US military to relish a qualitative edge over rivals.

However, in the last three decades, there has been a shift in the innovation environment from public to private-sector-led innovation. The change has been largely driven by the expansion of commercial market for technology. Companies started focusing on commercial innovation rather than on national security, and the DOD "has gone from being a producer to a consumer of innovation" (Lewis, 2021). Moreover, the US investment in scientific research and technology (R&D) has steadily declined since the 1970s (Independent Task Force Report No. 77, 2019). From reaching its peak at above 2 percent of gross domestic product (GDP) in the 1970s, US R&D was at about 1 percent in 2001 and 0.7 percent in 2018. Meanwhile, China has been continuously increasing its R&D expenditures. Since 2000, China has increased its scientific research and technology funding by an average of 18 percent annually. These challenges posed by China prompted The Task Force Report from 2019 to conclude that "China is closing the gap with the United States and will soon be one of the leading powers in emerging technologies" (p. vi).

Higher Education as the Primary Engine for the US Innovation

Although innovation in the US is still relatively strong, it has shifted to the private sector, and it focuses on commercial markets and profit rather than national security. For example, Apple—a company that once made strong investments in research and development—spent nearly $1 trillion on share buybacks from 2012 to 2018 (Medeiros, 2019). Pfizer, a pharmaceutical giant, followed a similar course of action and spent $139 billion on share buybacks. Between 2003 and 2013, many of the public companies in the S&P 500 index spent over half of their earnings to buy back their shares to boost stock prices (Lazonick & Mazzucato, 2013; Lazonick, Mazzucato, & Tulum, 2013). Medeiros (2019) makes a compelling argument that this money was disinvested from the companies' research, development and training for workers which supports the idea of the shift that the U.S innovation has experienced – from public to private and profit driven sector. There is, however, at least one specific establishment that could mitigate these challenges, reestablish the U.S government as a world leader in innovation, and encourage talent to participate in national security innovation strategy. This establishment is the higher education system. Higher education is the primary engine for US innovation due to the highly productive assemblage of talent and resources in its unique assortment of autonomous public and private universities (Lanford & Tierney, 2022). Through meaningful and sustainable partnerships between the US government and the nation's colleges and universities, we would fulfill our creative and innovative potential for the well-being of our country and its citizenry. This is the motivation behind the concept of *national innovation networks or national innovation ecosystems* (Janeway, 2012). The idea that educational institutions should form deeper partnerships with the government to deliver innovative and high-quality products, such as AI, advanced semiconductors, genomics, 5G technologies, and robotics, is in the best interest of both parties (The Independent Task Force Report No. 77, 2019). For at least the past several decades, many significant innovations have been produced outside of the defense economy (Gertner, 2013; Hiltzik, 2000). At the same time, public funding for higher education has decreased dramatically, especially since the economic downturn in 2008 (Mitchell, Leachman, & Saenz, 2019; Mitchell, Leachman, Masterson & Waxman, 2018;). Including the higher education system within the national innovation ecosystem would allow the DOD to link innovation back to national security and provide greater financial stability for researchers who need a long-term, interdisciplinary approach to developing innovations that can not only support US defense, but also solve some of society's most complex problems. Contrary to popular belief, universities are engines for innovation and entrepreneurship because they have disciplinary experts who have been trained in Ph.D. programs to take risks in their applied and theoretical research and have an entrepreneurial perspective in turning their research into products and services (Mazzucato, 2013; Mendoza, 2007; Rhoades & Stensaker, 2017; Tierney & Lanford, 2016a). However, such an approach would require the government to also increase their R&D expenditures, including federal and state strategic investments in universities.

The type of talent that the higher education system ought to provide for the national innovation ecosystem should also embrace diverse skillsets and backgrounds to meet the diverse challenges of constant and unpredictable change that is now the norm in economic, geopolitical, and military spheres. Since research has shown that new areas of knowledge, problem solving, and innovation are developed through social interactions (Amabile et al., 1996), then increasing the number of diverse faculty would increase our ability to meet these goals. As explained earlier in this paper, one way to boost the population of diverse thinkers is through equitable faculty socialization processes. Effective talent management starts with adequately developing a diverse body of graduate students to quickly become independent and fully functional members of the professoriate once they join their new workplace. This entails paying attention to individual and organizational cultures and making sure that individual potential is fully developed and in alignment with the system. Moreover, a sum of diverse faculty experiences and backgrounds would provide a more comprehensive and inclusive innovation environment. Such environment would better serve the needs of the DOD. This is because our national innovation network must be able to confront the difficulty of volatile, uncertain, complex, and ambiguous environment in which the return of great power competition is no longer a threat, but a reality, and in which China is threating our position as the leading power in emerging technologies. Hence, we ought to promote a more networked approach to innovation where different actors' contributions are valued and included and aim at mutual and desired goals, such as national security and economic independence.

INNOVATION ECOSYSTEMS

Contemporary societal changes are often described as a transformation to a knowledge-based society where learning and knowledge production occurs is innovation ecosystems (Manyika et al., 2013). Cai, Ma, and Chen (2020) explain that the core elements of innovation ecosystem are "increasingly interdependent and bind together by co-evolution/co-creation mechanisms, comparable to complicated relations among organisms is a bio-system" (p. 1). One aspect of an innovation ecosystem that makes it distinct from a business ecosystem or innovation system is *value co-creation*. Value co-creation refers to an interactive process of creating mutual goals, objectives, and utilities between producers and consumers (Smorodinskaya et al, 2017). Value co-creation can apply to a more fundamental or lower-order unit, such as university, and take place between faculty and organization, or it can apply to a higher-order unit, such as society, and take place between university and government. In such innovation ecosystems, knowledge is also context-dependent, rather than tactical or codified. Hence, as described previously, knowledge production transpires more often through social interactions supported by organizations that facilitate an environment of creativity and innovation, rather than in isolation (Tierney & Lanford, 2016b).

An environment of innovation further requires shared values that facilitate communication between individuals so that vital feedback from disciplinary experts and impacted communities informs research and development. These shared values align seamlessly with equitable socialization. As mentioned in the previous sections of this paper, a fairly executed socialization process embraces multiculturalism and promotes cultural responsiveness. This is critical to create a diverse context in which innovation can strive. Because of the bidirectional nature of the process itself—that is, members adapting to organizations and organizations adapting to their members—the best way to marry faculty and academic organizations, as well as academic organization and government, is through establishing a shared vision, mutual goals, and core values. This is consistent with the idea that culture shapes and is shaped by social interactions and social networks (Geertz, 1973). Moreover, this allows for codependent cultural forces, such as national and regional cultures, institutional cultures, and cultures based on individual identities to coexist in harmony. Thus, in a knowledge-based society in which both universities and the government are a part of innovation ecosystem, the roles of both entities should evolve to be more culturally, socially, and nationally responsible. In other words, global societal changes demand the broader roles of universities and governments within any given system of teaching, learning, and innovating.

National Innovation Network

A critical part of creating a *national innovation network* would be reconciling two intricately interrelated transformations: societal and university. In addition, a national innovation network would also require the support and pursuit of the idea of *co-innovation*, that is "the dynamically intertwined process of co-operation, co-evolution, and co-specialization within and across regional and sectoral innovation ecosystems (Carayannis et al., 2018, p. 153). Co-innovation is characterized by "collaboration, coordination, co-creation, convergence" (Saragih, 2018, p. 361), and its prerequisite requires the establishment of mutual—or at least complementary—goals and objectives, and that is the process of value co-creation for the transformation of the society.

In recent years, the European University Association (EUA) published a report in which it specified four roles for colleges and universities in regional innovation systems: 1) "Education: providing human capital for innovation", 2) "Research: knowledge co-production for private and public value creation", 3) "Knowledge exchange for innovation systems: from technology transfer to multi-actor co-creation", 4) "Strategic transformation: embedding innovation" (Reichert, 2019, p. 22-47). This specification entails how universities must take ownership of the following three duties to function in innovation ecosystems. First, the role of universities should change from merely knowledge and technology transfer to knowledge and technology exchange. Knowledge and technology exchange entails collective and collaborative learning between organizations from different sectors, both private and public, and it also entails value co-creation, where universities are enablers of generating shared values, not just creators of values. Second, universities should explore a new function of trust-building between different agents in innovation ecosystems. Cai, Ma, and Chen (2020) write, "the interactions among actors in an innovation ecosystem can be understood as social relations and the knowledge exchange is an outcome of social relations. (...) trust is considered a key factor to successful knowledge exchange and co-innovation" (p. 4). And third, universities should be both institutional entrepreneurs in the innovation ecosystem. Battilana et al. (2009) identify institutional entrepreneurs as organizations or agents that instigate diverse changes and actively participate in their implementation. This further highlights the responsible and responsive role that universities started to undertake more frequently while participating in innovation systems.

The EUA specification concerning the new roles that universities should undertake in order to fully and responsibly participate in democratic societies, could be adopted in the US to establish effective national innovation ecosystems and by this link innovation back to national security. Our nation's innovation ecosystems, built on a foundation of trust and responsibility, would require the establishment of longer-term partnerships between the government and the universities than the current system of short-term grants which all-too-often result in a lack of innovation due to incomplete and unsatisfactory outcomes. Additionally, each actor would have to contribute its fair share of people, spaces, and resources in order to co-create domestic values and purse national security goals. This is achievable under three conditions:

1) the government increases their R&D expenditures and funds specific projects at universities' labs, 2) the universities provide a diverse group of talented thinkers and researchers to fuel national innovation, and 3) the talented thinkers and researchers are first equitably socialized so that disciplinary expertise, multicultural perspectives, and entire segments of society are never overlooked. In short, equitable faculty socialization is a precursor for diversifying academic landscape that is required to breed innovation and meet the challenges of today's irregular academic, economic, political, and military landscape. Furthermore, innovation can be linked back to national security via establishing national innovation ecosystems, or networks, in which the roles of the government and the higher education system are defined by responsibility, trust, and codependence for the greater good of the nation and its citizens.

Conclusion

In summary, equitable faculty socialization in higher education systems is of the outmost importance to equip the government with diverse talent via national innovation ecosystems. Skillful preparation of more women and minorities to join the professoriate would diversify the academic workplace and increase opportunities for the United States to innovate and produce diverse knowledge that emerges from social interactions. More equitable socialization would also increase cognitive diversity and usefully disrupt academic workplaces, encouraging a landscape of constructive feedback and critique necessary for the refinement of ideas and theories. This environment, in turn, would diversify social interactions and the contexts in which knowledge is produced and thus increase our potential for innovation. Recently, general Michael X. Garrett from US Army described diversity as key American strategic asset (Garrett, 2021). The General explained that diversity and inclusion keep the US force strong, and it is what distinguishes our nation on the global stage. I would further argue that specifically diversity of thought, which is predicated in part on said demographic diversity, is our country's inherent advantage that we should use to advance national security, increase the qualitative edge over military rivals, and maintain economic independence. In short, innovation through diversity is essential to better meet today's security challenges.

REFERENCES

Amabile, T. M., Conti, R., Coon, H., Lazenby, J., & Herron, M. (1996). Assessing the work environment for creativity. *Academic of Management Journal, 39*(5), 1154-1184.

Asante, M.K. (1996). Multiculturalism and the academy. *Academe, 82*(3), 20-23. https://doi.org/10.2307/40251475

Banks, J. A. (1995a). Multicultural education: Historical development, dimensions, and practice. In J. A. Banks & C. A. M. Banks (Eds.). *Handbook of Research on Multicultural Education* (pp. 3-24). New York: Macmillan.

Banks, J. A. (1997). Multicultural education: Characteristics and goals. In J. A. Banks & C. A. M. Banks, (Eds.). *Multicultural Education: Issues and Perspectives* (3rd ed., pp. 3-31). Boston: Allyn and Bacon.

Barber, P. H., Hayes, T. B., Johnson, T. L., Marquez-Magana, L. (2020). Systemic racism in higher education. *Science, 369*(6510), 1440-1441. DOI: 10.1126/science.abd7140

Battilana, J., Leca, B., Boxenbaum, E. (2009). How actors change institutions: Towards a theory of institutional entrepreneurship. *Academy of Management Annals, 3(1),* 65–107. https://doi.org/10.5465/19416520903053598

Bauer, T. N., Bodner, T., Erdogan, B., Truxillo, D. M., & Tucker, J. S. (2007). Newcomer adjustment during organizational socialization: A meta-analytic review of antecedents, outcomes, and methods. *Journal of Applied Psychology, 92*(3), 707-721.

http://dx.doi.org/10.1037/0021-9010.92.3.707

Beebee, H. (2013). Women and deviance in philosophy. In K. Hutchison & F. Jenkins (Eds.), *Women in philosophy: What needs to change?* (pp. 61–80). Oxford University Press. https://doi.org/10.1093/acprof:oso/9780199325603.003.0004

Blau, F. D., & Kahn, L. M. (2017). The gender wage gap: Extent, trends, and explanations. *Journal of Economic Literature, 55*(3), 789–865. https://doi.org/10.1257/jel.20160995

Bogacz, M. T. (2021). *Gender Parity in American Academic Philosophy: A Promising Practice Study* (Unique identifier: UC15622688) [Doctoral dissertation, University of Southern California]. USC Digital Library. https://digitallibrary.usc.edu/Share/85t6dciey3u58e8qiiey3od464w85prh

Boice, R. (1991). New faculty as teachers. *Journal of Higher Education, 62*(2), 150-173.

Boice, R. (1992). Lessons learned about mentoring. In C. M. Wehlburg (Ed.)., *New Directions for Teaching and Learning, 50,* 51-61. Jossey-Bass. https://doi.org/10.1002/tl.37219925007

Cai, Y., Ma, J., & Chen, Q. (2020). Higher education in innovation ecosystems. *Sustainability, 12*(11), 4376. https://doi.org/10.3390/su12114376

Carayannis, E.G. Grigoroudis, E., Campbell, D.F.J., Meissner, D. Stamati, D. (2018). The ecosystem as helix: An exploratory theory-building study of regional co-opetitive entrepreneurial ecosystems as Quadruple/Quintuple Helix Innovation Models. *R&D Management. 48,* 148–162.

Cheryan, S., Ziegler, S., Montoya, A., & Jiang, L. (2017). Why are some STEM fields more gender balanced than others? *Psychological Bulletin, 143*(1), 1–35. https://doi.org/10.1037/bul0000052

Council on Foreign Relations. (2019, September). Innovation and national security: Keeping our edge. Independent Task Force No. 77. https://www.cfr.org/report/keeping-our-edge/

Crepeau, E. B., & Thibodaux, D. P., & Parham, D. (1999). Academic juggling act: Beginning and sustaining an academic career. American Journal of Occupational Therapy, 53, 25- 30.

Eagan, J. L. (2022, September 20). Multiculturalism. Encyclopedia Britannica. https://www.britannica.com/topic/multiculturalism

Education Encyclopedia. (Accessed July 2022). Multicultural education: The history, the dimensions of multicultural education, evidence of the effectiveness of multicultural education. https://education.stateuniversity.com/pages/2252/Multicultural-Education.html

Garrett, M. X. (2021 May-June). Military diversity: A key American strategic asset. *Military Review.*

Geertz, C. (1973). *The interpretation of cultures*. Basic Books.

Gertner, J. (2013). *The idea factory: Bell labs and the great age of American innovation*. Penguin.

Hiltzik, M. A. (2000). *Dealers of lightning: Xerox parc and the dawn of the computer age*. Harper Collins.

Hutchison, K., & Jenkins, F. (Eds.). (2013). *Women in philosophy: What needs to change?* Oxford University Press. https://doi.org/10.1093/acprof:oso/9780199325603.001.0001

Janeway, W. H. (2012). *Doing capitalism in the innovation economy: Markets, speculation, and the state.* Cambridge University Press.

Johnson, S. D., Lucero, C. (2003). Transforming the academic workplace: socializing underrepresented minorities into faculty life. In Marye Anne Fox (Ed.), *Pan-organizational summit on the U.S. science and engineering workforce: Meeting summary*. National Academies Press.

Kahn, S., & Ginther, D. (2017). Women and STEM. *IDEAS Working Paper Series from RePEc*. http://search.proquest.com/docview/1913082625/

Kelly, K., & Grant, L. (2012). Penalties and premiums: The impact of gender, marriage, and parenthood on faculty salaries in science, engineering and mathematics (SEM) and non-SEM fields. *Social Studies of Science, 42*(6), 869–896. https://doi.org/10.1177/0306312712457111

Kelly, McCann, K., & Porter, K. (2018). White women's faculty socialization: Persisting within and against a gendered tenure system. *Review of Higher Education, 41*(4), 523–547. https://doi.org/10.1353/rhe.2018.0024

Korte, R. F. (2007). The socialization of newcomers into organizations: Integrating learning and social exchange processes (ED504550). ERIC. https://files.eric.ed.gov/fulltext/ED504550.pdf

Lanford, M., & Tierney W. G. (2022). *Creating a culture of mindful innovation in higher education.* SUNY Press.

Lazonick, W., Mazzucato, M. (2013) The risk-reward nexus in the innovation-inequality relationship: Who takes the risks? Who gets the reward? Industrial and Corporate Change, 22(4), 1093-1128. https://doi.org/10.1093/icc/dtt019

Lazonick, W., Mazzucato, M., & Tulum, O. (2013) Apple's changing business model: What should the world's richest company do with all those profits? Accounting Forum, 37(4) 249-267. https://doi.org/10.1093/icc/dtt019

Liera, R. (2019, February). Implementing racial equity in faculty hiring through inquiry. In R. Johnson, U. Anya, & L. Garces (Chairs), *Envisioning racial equity on college campuses: Bridging research-to- practice gaps for institutional transformation.* State College, PA.

Leuschner, A. (2015) Social exclusion in academia through biases in methodological quality evaluation: On the situation of women in science and philosophy. *Studies in History and Philosophy of Science Part A, 54,* 56-63. https://doi.org/10.1016/j.shpsa.2015.08.017

Lewis, J. A. (2021, April). *Linking national security and innovation: Part 1.* Center for Strategic and International Studies. https://www.csis.org/analysis/linking-national-security-and-innovation-part-1

Lewis, J.A. (2021, October). *National security and the innovation ecosystem.* Center for Strategic and International Studies. https://www.csis.org/analysis/national-security-and-innovation-ecosystem

Manyika, J., Chui, M., Bughin, J., Dobbs, R., Bisson, P., & Marra, A. (2013). *Disruptive technologies: Advances that will transform life, business, and the global economy.* McKinsey Global Institute.

Mazzucato, M. (2013). *The entrepreneurial state: Debunking public vs. private in innovation.* Anthem Press.

Medeiros, J. (2019, August 10). This economist has a plan to fix capitalism. It's time we all listened. *Wired.*

Mendoza, P. (2007). Academic capitalism and doctoral student socialization: A case study. *Journal of Higher Education, 78*(1), 71-96.

Mitchell, M., Leachman, M., Masterson, K., & Waxman, S. (2018). *Unkept promises: States cuts to higher education threaten access and equity.* Center on Budget and Policy Priorities. https://www.cbpp.org/sites/default/files/atoms/files/10-4-18sfp.pdf

Mitchell, M., Leachman, M., & Saenz, M. (2019). State higher education funding cuts have pushes costs to students, worsened inequality. Center on Budget and Policy Priorities. https://www.cbpp.org/sites/default/files/atoms/files/10-24-19sfp.pdf

Mitchell, R., Nicholas, S., Boyle, B. (2009). The role of openness to cognitive diversity and group processes in knowledge creation. *Small Group Research, 40*(5), 535-554. https://doi.org/10.1177/1046496409338302

National Academy of Sciences. (2000). *Who will do the science of the future?* National Academy Press.

National Center for Education Statistics. (2022). Characteristics of postsecondary faculty. Condition of Education. U.D. Department of Education, Institute of Education Sciences.

Retrieved July 10, 2022, from https://nces.ed.gov/programs/coe/indicator/csc

National Research Council. (1997). *Enhancing organizational performance.* Washington DC: The National Academy Press. https://doi.org/10.17226/5128.

National Science Foundation. (2014a). *Integrated postsecondary education data system, 2013, completions survey.* National Center for Science and Engineering Statistics.

National Science Foundation. (2019). *Doctorate recipients, by subfield of study and sex: 2018.* National Center for Science and Engineering Statistics.

Posselt, J. R. (2016). *Inside graduate admissions: Merit, diversity, and faculty gatekeeping.* Harvard University Press.

Posselt, J. R. (2020). *Equity in science: Representation, culture, and the dynamics of change in graduate education.* Stanford University Press.

Reichert, S. (2019). *The Role of universities in regional innovation ecosystems;* EUA.

Rhoades, G., & Stensaker, B. (2017). Bringing organizations and systems back together: Extending Clark's entrepreneurial university. *Higher Education Quarterly, 71*(2), 129-140.

Saragih, H. S., Tan, J. D. (2018). Co-innovation: A review and conceptual framework. *International Journal of Business Innovation and Research, 17*(3), 361–377. DOI:10.1504/IJBIR.2018.095542

Shanghai Ranking. (2021). *2021 Academic ranking of world universities.* https://www.shanghairanking.com/rankings/arwu/2021

Slay, K. E., Reyes, K. A., & Posselt, J. R. (2019). Bait and switch. Representation, climate, and tensions of diversity work in graduate education. *The Review of Higher Education, 42*(5), 255-286. http://doi.org/10.1353/rhe.2019.0052

Smorodinskaya, N., Russell, M., Katukov, D., & Still, K. (2017,

January 3-7). Innovation ecosystems vs. innovation systems in terms of collaboration and co-creation of value. In *Proceedings of the Hawaii International Conference on System Sciences*, Honolulu, HI, United States.

Song, S. (2020). Multiculturalism. In Edward N. Zalta (Ed.) *The Stanford Encyclopedia of Philosophy*, Fall 2020 Edition. URL = <https://plato.stanford.edu/archives/fall2020/entries/multiculturalism/

Stewart, A. J., & Valian, V. (2018). *An inclusive academy: Achieving diversity and excellence.* MIT Press. https://doi.org/10.7551/mitpress/9766.001.0001

Thomas, K. M., Willis, L. A., & Davis, J. (2007). Mentoring minority graduate students: Issues and strategies for institutions, faculty, and students. *Equal Opportunities International, 26*(3), 178-192. DOI:10.1108/02610150710735471

Tierney, W. G., & Lanford, M. (2016). Conceptualizing innovation in higher education. *Higher Education: Handbook of Theory and Research, 31,* 1-40. https://doi.org/10.1007/978-3-319-26829-3_1

Tierney, W. G., & Lanford, M. (2016b). Creativity and innovation in the twenty-first century university. In J. M. Case & J. Huisman (Eds.), *Researching higher education: International perspectives on theory, policy, and practice* (pp. 61-79). Society for Research into Higher Education. Routledge

Tierney, W. G., & Rhoads, R. A. (1993). Enhancing promotion, tenure and beyond: Faculty socialization as a cultural process. *ASHE-ERIC Higher Education Report No. 6.* School of Education and Human Development. George Washington University, Washington, DC.

U.S. News and World Report. (2022). *2022 Best global universities ranking.* https://www.usnews.com/education/best-global-universities/rankings

Villarreal, C. D., Liera, R., & Malcom-Piqueux, L. (2019). The role of niceness in silencing racially minoritized faculty. In A. E.

Castagno (Ed). *The price of nice: How good intentions maintain educational inequity* (pp. 127-144). Minneapolis, MN: University of Minnesota Press.

Zambrana, R. E., Espino, M. M, Castro, C., Cohen, B. D. & Eliason, J. (2014). American Educational Research Journal, 52(1), 40-72. DOI:10.3102/0002831214563063

5

SYMPOSIUM TRANSCRIPT
DR. MARGARET E. KOSAL

Dr. Margaret E. Kosal

As presented at the 2022 Strategic and Security Studies Symposium
Hosted by the Institute for Leadership and Strategic Studies
University of North Georgia

Heath Williams:
Good afternoon. I'm Heath Williams. I'm the Director and Federal Liaison here at the University of North Georgia, and it gives me great pleasure at this time to introduce our next keynote speaker, and that is Dr. Margaret Kosal. Dr. Margaret Kosal is an associate professor in the Sam Nunn School of International Affairs at the Georgia Tech Institute of Technology. Her research explores the relationships among technology, strategy, and governance. Her research focuses on two often intersecting areas, which are reducing the threat of weapons of mass destruction and understanding the role of emerging technologies for security. During academic year 2016 through 2017, Dr. Kosal served as a senior adjunct scholar to the Modern War Institute at West Point. She previously served as a senior advisor to the Chief of Staff of the United States Army, as a science and technology advisor within the Office of the Secretary of Defense, and as an associate to the National Intelligence Council. Dr. Kosal is the recipient of multiple awards, including The Office of the Secretary of Defense Award for Excellence. Dr. Kosal is the author of *Nanotechnology for Chemical and Biological Defense*, published by Springer Academic

Publishers 2009, which explores scenarios and strategies regarding the benefits and potential proliferation threats of nanotechnology and other emerging sciences for international security. Her other books include *Weapons Technology Proliferation: Diplomatic, Information, Military, Economic Approaches to Technological Proliferation* and *Technology and the Intelligence Community: Challenges and Advances for the 21st Century*. Dr. Kosal's academic degrees include a PhD in chemistry from the University of Illinois at Urbana Champaign and a Bachelor of Science in chemistry from the University of Southern California. At this time, I would like to give the floor to Dr. Maggie Kosal.

Dr. Margaret Kosal:

Thank you. I'm going to speak about a number of different topics.

If General Brown was still here, one of the questions I would have asked him would have been something like, how did you come to the conclusion that you stated so strongly that government no longer plays a role in innovation? Because, if you notice, I started out with a PhD in chemistry. Before I finished that, I had, along with three colleagues, started a high-tech startup company. And this is the model that we see more and more. Of course, Google wants your technology; the other big companies want the technology, but essentially these startups that spin out of universities like Georgia Tech, that's the place where you can have the risk and where you can have the opportunities to fail that was talked about earlier in the symposium. We don't have a Bell Labs anymore. So, ignoring the critical role of universities and government funding, a significant amount of which comes from the DOD, by the way, that's downplaying the role of universities. That also does not take into account the role of universities today in our innovation infrastructure and innovation ecosystem, which means that you don't get the full picture.

As was mentioned, I've done a few different things inside and outside of the government. *Nanotech* was my first book; this book came out of the *Disruptive Game Changing Technologies* that came out in I think 2020. That one was most recently recognized by NATO as its number two book for 2021. I'm number two; I've published a couple other volumes as well. Within the Department of Defense, I have interacted in a variety of different roles usually at some place bridging my scientific background with international affairs or political science. And one of the most important things that I've come to perceive and to appreciate is how OSD does not reflect the services. And a whole lot of academics who are doing international affairs look to OSD, the Office of the Secretary of Defense, and think that reflects the services. I'm smart enough now to recognize that even being an HQDA does not really reflect the services, but at least you're getting a little bit closer, and I speak to that because, right now, I'm serving on these National Academy of Sciences committees and trying to make sure that we're bringing in the perspective from SOCOM, the perspective from the 20th CBRNE, because these are two studies that are being funded by the Department of Defense.

On to what I'm going to talk about today. I'm going to start with the strategic context. I'm going to talk about three different emerging technologies, and I want to emphasize that I'm talking about emerging technologies not emerged technology. I am talking about things that are really just coming to the forefront, but we're not going to be talking about just the technologies. We're talking about them in the context of the geopolitics, such as who's going to be the first adopter, the PRC or the United States? And this is some research that literally was just published about a month ago.[1] And then some nanotechnology in the form of biometric materials.

---

1 Kosal, M.E., and J. Putney, "Neurotechnology and International Security: Predicting Commercial and Military Adoption of Brain-Computer Interface (BCI) in the US and China," Politics and the Life Sciences, 2022, pp 1-23, https://www.doi.org/10.1017/pls.2022.2

I want to start off again putting this in context. This is not my research; this is research from Peter Turchin and colleagues in which they looked historically at the emergence of groups of technologies that had significance in conflict.[2] What you see is the data; this is where groups of technology emerged historically that had significant effect in in conflict. We see where civilization emerged, large-scale polities is the technical jargon there. And then this was their model, so they're looking at this interaction between technology and the emergence of civilization. And I like to start off with this because, particularly those of us who are involved in security, national security, we're often doing threat assessments, for good reasons. You have to hedge, but we also need to be cognizant that technology can be a really good thing. Even technology that has implications for conflict, because we see a correlation between groups of technologies used in conflict and the emergence of civilization—unless you want to argue that the emergence of civilization was the first problem, and we should have stopped there. I know there are some that do make that argument.

I'm interested in technology's role in politics and war, and, just as importantly the role of politics and war and technology. We are not technologically deterministic here. Here's a quote from 2010 NATO new strategic concept, the first major review after the collapse of the Soviet Union (they are undergoing a new review right now).

"Less predictable is the possibility that research breakthroughs will transform the technological battlefield. Allies and partners should be alert for potentially disruptive developments in such dynamic areas as information and communications technology, cognitive and biological sciences, robotics, and nanotechnology.

"The most destructive periods of history tend to be those when the means of aggression have gained the upper hand in the art of

---

2   Turchin, P., T.E. Currie, E.A.L. Turner, and S. Gavrilets. 2013. "War, Space, and the Evolution of Old World Complex Societies." *PNAS* 110: 16384–16389.

waging war."[3]

So highlighted in that quote are a number of technologies that were asserted to be potentially significant in the nature and character of the future of warfare. This is the long list of the security puzzles and the kind of things that I work on in my research group. I'm not going to go through all of these, but I want to highlight a few because, ultimately, we're thinking about national security and higher education. These are the kind of questions that my students, that other students who are in international affairs and in political science, that we're trying to grapple with, to understand the causal relationships. And this is often much harder than the work I did when I was a PhD chemist. I'm still a PhD chemist; I just don't work in the lab anymore. We can control experimentally a lot more in the physical sciences. If I tried to do an experiment in which I was looking at proliferation, I'd get the FBI at my door or somebody else. I don't have another planet.

Looking at a number of these different questions in terms of the work that I'm doing, I want to look at things like, do these technologies, these emerging technologies, have a strategic value? Or are they just evolutionary, and they're contributing? Is there something truly revolutionary about them? Because things that are truly revolutionary can change the character and conflict of war. We're looking at things like, are their technical aspects, are there structural pieces, and are there political pieces? And ideational again, not just looking at any one aspect.

I've spent a lot of time, depending on sort of what's popping in the news, dealing with hope or horror scenarios. A fair bit of this is sort of calming people down: no, we don't need a treaty for x, and here's why. And then, ultimately, I'm interested in things that can be implementable. How do we enable humans to benefit from our own creativity, while trying to reduce the risk or minimize the risk from misuse.

---

3   https://www.nato.int/cps/en/natohq/topics_82705.htm

So, AI and geopolitics; I'm going to look at a couple different things within these statements and rhetoric and then ask this question: who has the advantage in an AI battle? I don't have the answer, but I have more questions. So, what is AI? There is no single definition of artificial intelligence. I very much like this definition; this was from the NDAA.

Military applications of artificial intelligence are an area that we can spend a very long period of time talking about these. These are some of the things that one is anticipating and is already hearing about. A lot of them have to do with dealing with the amazing amounts of data and being able to make sense of that data because humans just can't do it. But there also are some AI applications like the SharkSeer Program. Every email that comes into a military address goes through this because it's looking for spyware and for other types of malwares, and it catches it faster than previous ways of doing it that were human dependent. This initially was an NSA project that got opened up during a Congressional hearing when someone mentioned it.

I really want to focus more on is looking at again what are these types of AI right now. What we have is narrow AI. General AI, at this point, is still science fiction that of the sort of super intelligent, that the machines are smarter than humans. If we start digging into the geopolitics and the discourse, because discourse in geopolitics does matter, here we have a statement from September 2017 from Vladimir Putin.

"Artificial intelligence is the future, not only for Russia, but for all humankind. It comes with colossal opportunities, but also threats that are difficult to predict. Whoever becomes the leader in this sphere will become the ruler of the world."[4]

This is a comment that Putin made, and, notably, he made this assertion during a speech that was televised to all the primary and secondary school children in Russia. It was Russia's Children's Day,

---

4 https://www.rt.com/news/401731-ai-rule-world-putin/

and he's saying that whoever has AI is going to become the ruler of the world. There is implicit security in his comments, there's implicit prestige, and we're also talking about directly saying things that are part of indoctrination.

Now of course there are also folks in the United States and other places who were listening to this as well. When we hear about AI or frankly any emerging technology, it is super important to be cognizant of hype. Okay, [References the showing CatGenie automatic cat litter ad featuring the "Cat-Genie A.I."], this is the greatest cat litter box in the world. You do not have to do anything, but recently, they are advertising it with the internet of things; we're going to have artificial intelligence in your cat litter box. Now probably by that definition of narrow AI there's a machine learning algorithm that is in there, but again we have to be cognizant of hype, and we have to be cognizant of the role of discourse, particularly when it is coming from political leaders.

Another example, that came out in 2017. This is Russia's, at the time, it was called status 6; it is thought that this was an intentional leak. You will notice that this is an image that was taken over the shoulder of an individual in the Russian military. So here we have PowerPoint capabilities that then became assumptions about actual capabilities that the Russians might have now. Again, we have to be cognizant of using the narrative. The Russians have a deep history going back to at least the 1700s of Maskirovka and some of these other activities. Just because Russia claims they can do something does not mean they can do something. And it isn't just in AI. This is from the first decade of the 2000s; these are a variety of statements, some by Putin, some by Medidata, some by the person who was in the role of Shiogonow about nanotechnology. I call these statements 'glory to nanotechnology.' We see again this rhetoric about emerging technologies, and a big part of this is about showing that you have technical capabilities that are commensurate or associated with prestige.

By the way thermobaric have gotten a whole lot of attention recently with accusations, and it does appear that they've been used within the Russians in Ukraine. Well, this image is of the FOBA, the father of all bombs, the largest non-nuclear bomb ever detonated. The Russians did this back again in the first decade, and they asserted that there was nanotechnology that enabled this thermobaric bomb. It's not clear because, of course, they don't give us the details, but there are a number of different observations that suggest that it wasn't quite as innovative as they were thinking about. And again, this is back to, remember, I talked about sort of those tests, some of those technical issues: is it changing something? Or is it just evolutionary? Likely they were using nanotechnology to increase the surface area of the propellants that were part of the explosives. That's not new; that's just making a better explosive. It matters, but it's not a strategic game-changing application of technology.

Moving on to the PRC's AI strategy, where we can look to see indicators about discourse. It came out in 2017. Typically, you see strategy documents where China is talking about the economy, talking about national capabilities. What was most notable about this strategy, this is one of the first times that China in a strategy associated with technology has explicitly called out national security. Most of the time in their strategies they practice an intentional ambiguity. If you look at their strategies with respect to life sciences, you can read into it, but it isn't called out explicitly. There is something different here. And we also see that nations do see AI as being significant with regard to national security, military applications, and conflict. We need to understand a lot better, what that means specifically, and again because we have to look at ourselves. [References the slide]

This is a quote from Admiral Jeremiah.

"Military applications of molecular manufacturing have even greater potential than nuclear weapons to radically change the balance of power."⁵

He had just recently retired from the Navy at that time. He said this quote well over 20 years ago: "military applications of molecular manufacturing"—just imagine substituting nanotechnology—"has even greater potential than nuclear weapons to radically change the balance of power." Is anybody here going to claim that nanotechnology has changed the balance of power? It hasn't; right now, the single area that nanotechnology is most significantly contributing to is health and beauty products, which from a capitalistic perspective a whole bunch of people are making a lot of money off of it, and it's doing some good things especially as—I'm a gen-Xer—we're starting to get older. But that's not strategically significant, and that's not changing the balance of power. I say this not to not to suggest that these things are not important but to remind us that we need to examine our own assumptions even when they're about ourselves.

Now we don't have anybody who makes the kind of rhetorical statements that Putin does, but again we do see, particularly in terms of emerging technologies, right now there's been a proliferation of a lot of statements on emerging technologies, a lot of statements that aren't necessarily backed up by good science, never mind good political science.

On to another organization. If you have not seen the Slaughterbots video, I encourage you to Google it on YouTube. It is an amazing piece of narrative. It looks almost like a Ted Talk, where you've got this scenario of an unnamed organization that has these miniaturized drones that can target human beings selectively. So, was this a prescient warning? Well, the intention of this organization that put this together is to influence the UN. They want to pass a treaty related to the prohibition of armed robots in conflict which, we're sort of like, "well, what does that mean?" We

---

5 * "Nanotechnology and Global Security," (Palo Alto, CA; Fourth Foresight Conference on Molecular Nanotechnology), November 1995

do need to be cognizant of the use of scientists. Here we have a computer scientist who is advocating for these prohibitions, and this is a great example of where scientists, their prestige, is being leveraged to push forward a political agenda about limiting the use of UAVs.

Now, one of the aspects of being somewhat a reformed scientist as I am, you might say, scientists love to come out and say we need a new treaty on x. Not recognizing that there exist aspects of the laws of armed conflict, aspects of the Geneva Convention, the Martins Clause, etc. that actually cover a lot of things. But there are states like Russia and like China that love it when the international discourse typically, the international legal discourse, gets muddied because, "Oh yeah we need to be thinking about, we need to devote our attention to x treaty to prevent nanotechnology, to prevent grey goo."

And again, I want to step back. Yes, the computer scientist in the Slaughterbots video is an incredibly smart person; he is a brilliant computer scientist; he is somebody who I am convinced his heart is in the right place. He wants to be an ethical American, but unfortunately this ultimately benefits states like China and Russia who don't want us to focus on the misuse of drones for targeting civilians. And we don't need a new international treaty to say it's illegal to indiscriminately target civilians. We've got that; we don't need a new treaty.

So, sort of wrapping up on the AI section, if you are interested in artificial intelligence for implications with national security, I highly recommend Greg Allen and Tan Chan's piece.[6] It's a few years old now, but this is probably the best piece in terms of a sort of overarching addressing of AI. One of the things they write about are these potential transformative scenarios, and their very first one is that "supercharged surveillance through AI is going to bring about the end of guerrilla warfare." This came out in

---

6   Greg Allen & Taniel Chan, *Artificial Intelligence and National Security,* Harvard Belfer Center, July 2017

2017, and, by the way, Greg Allen went on to the Joint Artificial Intelligence Center (JAIC) in OSD. The office that is supposed to be the coordinating piece across the DoD. Again, these are super smart people; these are good Americans. They say this is the most likely scenario. And I said, "Well hold on, let's think about creative countermeasures." I suspect that there are a few of you who come from the unconventional warfare side, who have thought about creative countermeasures. So have the folks protesting in Hong Kong, before everything got shut down. We saw mainland China attempting to use machine learning algorithms to identify individuals. What the protesters used were green laser pointers to confuse the optics. They also used umbrellas because if you can't get a good image of the person, you can't identify them. They used mylar blankets which are actually a remnant of the SARS epidemic—before the most recent COVID epidemic—to avoid detection by infrared (IR). Here we have some incredible countermeasures, and, again, these are the protesters in Hong Kong. These are not trained guerrillas. The assumption about the directionality of emerging technologies, that it is going to benefit these surveillance states so that guerrilla warfare is not possible, well, I think we need to rethink that. We are seeing that is it still necessary to seize and hold territory. We can invoke Carl Von Clausewitz. In terms of what we're seeing going on in Ukraine, you've got to hold territory.

Okay moving forward. Neurotech again. I'm going to talk very briefly about the global picture on neurotechnology, the cognitive neurosciences, brain science. I'm going to talk about different types of brain computer interface. And then I'm going to talk about, how do we think about who might be the first adopter: the US? Before diving into our result, why might you want to study cognitive science and neurotech from a national security perspective? There are a whole lot of reasons. We might have advantages for ourselves and our allies, or we might need to counter something that an adversary is doing. Right now, I'm involved in a group, a NATO

MCDC group, that is looking at human performance enhancement, and part of that's augmentation, part of its degradation. Can neurotechnology be used to destroy the cognitive capabilities? And then counter measures, which fields are the ones that we need to be looking at? It's not like the 1930s and 1940s, when you could just look at nuclear physicists— and by the way even when we were developing nuclear weapons, you still had to look at the chemists because they were part of it too, and the engineers, you can never forget the engineers. Cognitive science is shedding light on new deterrence approaches deterrence, because deterrence is based on assumptions of rationality among actors. Discoveries in the cognitive neurosciences are pointing out that some of those assumptions of rationality may not hold. And then implications for information operations, command and control, etc., if we start implementing some of this, what does it mean in terms of sustainment? And might we change the character of warfare?

My slide shows references to what is often referred to as "Havana Syndrome," which is an example of a degradation that has been asserted as potentially having been done by some sort of electronic or other capability to cause individuals, mostly members of the Department of State, to suffer illnesses. What exactly caused it? We aren't sure, and there only are about ten individuals right now we can't explain right away, but this got highlighted in the February ODNI report to Congress on the threats. The Havana Syndrome as it is conceived is an example of human performance degradation. Some people think Russia is doing it—again, there are a lot of questions we need answered to understand this better and to consider what it means for the nature of conflict.

In the recently published study, we looked at global investments in brain computer interface (BCI). Brain science is the thing right now that's competing with artificial intelligence for the most attention. You can see these very large investments across the globe. As one look at BCIs, they can be divided into something that is invasive, i.e., does it actually have to be surgically implanted? Or is it something that is on the surface? And then you can also look at it, does it just "read" your thoughts? Does it just take the thoughts, as measured by the neurochemical impulses, electrical impulses, and translate them? Is it read-only? Or can you read and write? Can it "read: what you're doing, but also write to your brain?

Another way of looking at this is, these are a variety of different applications some of which are being pursued actively, some of which are much more notional. You can see that all of them have good applications that are beneficial to society. Most of the work that DARPA has been funding, in terms of brain computer interfaces, has to do with prosthetics making better limbs that can function better for service members who have been injured. This is something that is really important to do. But often that also means that to be able to control a limb you also might be able to control a UAV with your brain. So, they are fundamentally dual use. And there at least three definitions of dual use that I'm aware of. These things can be used for civilian purposes; they can be used for military purposes, we also can look at them as things that can be used for beneficial purposes or misused against us.

Does the neurotechnology enable restoration or enhancement? Again, one can imagine a whole number of different medical applications. You can also imagine things where you want to monitor the stress of say a fighter pilot, or you might be looking at, can we influence what someone is thinking? What are their capabilities? The military has done a significant amount of research to prove that the most important thing to your cognitive ability is sleep. Cognitive ability is impaired by lack of sleep. We also have scientifically shown that the best thing to keep you awake is caffeine. You don't need any fancy drug, just caffeine. Is there a way to degrade that? You might have gotten a good nine hours of sleep—wouldn't that be wonderful—but we're going to subject you to something that is going to make it seem like you only had two hours.

Brain to brain communication is another thing: can we just "talk" telepathically? I'm going to talk to you right now: wouldn't that be useful? So those are the kind of questions that prompted the study by one of my former PhD students from quantitative biosciences and I. We wanted to investigate and understand who's likely to be the first adopter of brain computer interface technology? We leveraged work that had been previously done looking at other technologies, and total credit to my PhD student; she came up with an amazingly novel variable for us to look at. In this study, we looked at qualitative factors; we looked at semi-quantitative variables; and we looked at quantitative variables.

Another way to break them up is that we looked at variables that reflect political institutions, cultural variables, economic variables, and then some technical variables. To cut to the chase, if you assess the United States and the PRC based on a number of the different variables, the United States unquestionably has the better innovation system, and we might be putting more money at it, but because of some of the social and cultural norms and because of the specific regime type along with this novel variable of research monkeys, we're less likely than China to be the first adopter.

The research monkey variable is worth speaking a little more to. If you want to develop something that uses brain computer interface, you've got to test it in a primate. In the United States, at the start of the Covid epidemic, a great deal of our brain research was shut down because China stopped shipping over monkeys that are used in the testing. China already has the largest breeding colonies. By the way, this is even a more dire situation for Europe where, because of some EU regulations, working with primates is extremely restricted. Again, a novel variable; it doesn't tell the whole story, but a novel variable if you want to think about this emerging technology.

So, wrapping up here. Meta materials, are a type of nanotechnology to which I applied a threat assessment framework that was developed with a number of students. What are meta materials? Meta materials are biometric nano materials. The idea is that you are going to be able use meta materials to become invisible, transparent. Now that's cool; that's the science; that's the gee whiz.

What do meta materials potentially mean for security? A colleague and I went through and looked at it in the context of camouflage, which is not new. We've had camouflage for a long time, but is there something different about meta materials that are likely to change the security dilemma? This is an example of the type of threat assessment that we do, where we start off, where we've got a number of technical capabilities and challenges, then we looking at a variety of different operational environments. And we go back to assess how that new technology may affect conflict between the US and a number of different possible adversaries, such a near-pure contender, China, Russia, non-state actors, etc. What this analysis shows is that for many of the environments and many of the potential adversary types it's probably going to be a toss-up whether meta-materials benefit the defender or the challenger. For example, China is likely to have meta materials capabilities; they're probably not going to have an advantage over us, or it's going to be minimal in the context of our other capabilities. Blue is where we are likely to have an advantage; red are the ones where it is likely to benefit the adversary. And what you see there are situations where we're looking at intrusion into a border or non-state actors and that becomes this possibility that an adversary might have an advantage. If you think it's hard to find people coming across the border already, imagine they're invisible and you can't detect them with IR or anything else. This is the type of analysis that we do. In the actual publication, it was much more significant, but again I just want to highlight that, and we've done this for a number of different technologies.

In summary, I have given you a survey, quick and dirty, into emerging technology in the context of security implications, the geopolitical rhetoric, and the emphasis on it. And that phrase "emerging technology" might be the only thing proliferating faster than the technologies themselves. We see the rhetoric in a whole number of different places. We've got to be cognizant of dual use; that is going to be a challenge with any of these technologies. In conclusion, we need a more robust understanding of the security implications of emerging technologies. Technologies are likely to exacerbate existing social and political divides, aka, instability. Often, when we're thinking about emerging technologies, we don't think about it in that context; we think about it only in the high-tech context. The difference between beneficial and dangerous research is often only one of intent, and that makes it a lot harder than many previous technologies. You can use these things for good, for very good, including good for the economic well-being of the United States and our allies. Governance is a huge challenge; I'm not one who just throws my hands up. It is something that we need to address, but we need to recognize that, if we try to restrict only ourselves to minimize the risks, we may inadvertently harm ourselves and harm our allies because there are other places and other people that aren't going to impose restrictions. At the same time, this is the system; we've got we got to work with it; we need to foster proactive international scientific cooperation. The nuclear era might be seen as the height of track two diplomacy; we need to have a whole lot more attractive diplomacy; that's scientists to scientists or political scientists to political scientists. I'm at the Sam Nunn School of International Affairs. The Senator Sam Nun—we just call him the Senator—is still active.

And again, in the end, it's ultimately about the people. I would be remiss if I did not acknowledge some of the students, these variety of students. Some are Ph.D. students; some were masters students who worked with me from across Georgia Tech.

Finally, thank you for your kind attention.
[See Appendix for corresponding PowerPoint presentations.]

6

SCIENCE, TECHNOLOGY, AND STRATEGIC ANALYTICS

Panelists:
- Colonel, US Army (Retired) Eric Toler, Executive Director of the Georgia Cyber Center
- Greg Parlier, Ph.D., Colonel, US Army (Retired), Adjunct Professor of Operations Research at North Carolina State University
- Lieutenant Colonel Christopher Lowrance
- J. Sukarno Mertoguno, Ph.D.
- Dr. C. Anthony Pfaff

Moderator:
- Steven Weldon, Director Cyber Institute, School of Computers & Cyber Sciences, Augusta University

As presented at the 2022 Strategic and Security Studies Symposium
Hosted by the Institute for Leadership and Strategic Studies
University of North Georgia

Steven Weldon:

I want to welcome you to the best panel three that you will have all day so, I will certainly claim the best panel three moderator title. This is a very interesting panel, I think you're really going to enjoy it, and yes, we are the last graded evolution before the end of the day so let's go ahead and get started. To start off you know our panel three topics are science, technology, and strategic analytics, so this is an interesting topic but, to recalibrate ourselves to what got us here to panel three, our symposium theme is what is the current relationship between higher education and the military in the United States? And what is the future of that relationship? And then we heard from panel one discussing how higher education fills the security gap in the post-Cold War era, onto panel two rethinking higher education practice to stimulate innovation and global security, and then certainly Dr. Kosal's incredible keynote address on emerging technologies really sets the stage for what we are going to discuss today with this esteemed panel.

So, next, I'd like to introduce all to the entire panel, and then we will go as the previous panels, down through the speakers, and then open for questions. So, I'll start off with Eric Toler, who is Executive Director of the Georgia Cyber Center and a Colonel in the US Army retired. Dr. Greg Parlier is a professor of operations research at North Carolina State University, President of GH Parlier Consulting, and Colonel US Army retired. Dr. Sukarno Mertoguno is faculty at Georgia Tech, Georgia Institute of Technology School of Cyber Security and Privacy and there he is the Director for Cyber Systems Analysis, Formulation, and Automation and, of note he was a Department of Navy civilian at the Office of Naval Research. And then to my immediate left, I have Dr. Anthony Pfaff Colonel United States Army retired, and Dr. Christopher Lowrance Lieutenant Colonel promotable, congratulations, US Army, and both from the Army War College.

So in brief that is our panel for this afternoon and so now first up our lead-off batters, we have Dr. Anthony Pfaff and Dr. Christopher Lowrance gentlemen take it away.

Dr. Anthony Pfaff:

We're going to talk to you fairly quickly about a piece of what is really part of a larger project, and that larger project is about how to integrate artificial intelligence and data science into the military's professional expert knowledge. We're trying to answer the question, How do we trust not just this technology but the system in which this technology operates? And we kind of map that expert knowledge in terms of not just the technical but also the human talent management kind of concerns, the ethical concerns, and the political concerns and that political kind of mapped into both civil-military relations, as well as how different government stakeholders and acquisition stakeholders interact to make this an efficient process.

And what we're learning is, and we've covered this a little bit already today, is that integrating AI technologies poses a special challenge, and General Brown was getting at this earlier, but unlike previous arms races, like the race to the atomic bomb in World War II, expertise as he said rests largely with industry and academia, not with the Department of Defense. Moreover, it's not like we can just do another Manhattan project and build a small core of expertise within the military. Almost everyone will have to develop some level of AI and data literacy if the US military is to realize the full potential of these technologies.

Now to get there, to better understand this impact, our studies draw on lessons learned from Project Ridgeway, which some of you may know is an effort by the 18th Airborne Corps to draw on commercial and academic expertise to become AI-ready at the operational level. Moreover, this project is very much a ground-up effort, where the Corps engages in the private sector directly to take advantage of commercially available data and algorithms to support targeting in the deep fight.

Now, I'm going to go give it over to Chris who's going to talk about sort of the educational training challenges that we've experienced and where academia and, to a lesser extent, industry might fit in.

Dr. Christopher Lowrance:

In the past, we've had the luxury of having, generally speaking, technological superiority in the recent combats, especially in Iraq and Afghanistan. But if we're looking now forward to the Great

Power competition we're facing, if we have to go to war with a near-peer threat, that's going to be challenging and especially in the domain of AI. And AI itself is open science, generally speaking; the frameworks and the tools to build AI applications are widely available. So, to a general population, if you want to build a widget or an application that is AI-based you can go do so openly. Those kinds of frameworks to build neural network learning applications are available. So, with that kind of premise, how do we gain superiority? And that was actually a question earlier right. And our argument is we've got to turn to our people, just like we have in the past when we were facing the Soviet Union when we were going to be outnumbered, potentially outgunned and we knew it was going to come down to our people and the training of those folks and our military members. Similarly, we've got to turn to higher education to get the skill sets that we need to more effectively employ AI than our adversaries. And so that's really going to be the gist of our talk this afternoon.

So we can frame the discussion, I'm going to put this in the context of AI-based targeting. How do we incorporate artificial intelligence, more specifically, machine learning into the targeting process? Then I'm going to present some challenges and consider how do we potentially overcome those challenges? And then, obviously, our argument is to leverage education and training to get to where we want to be, stronger than our adversaries.

So, AI in the targeting process. As you all know, generally speaking, with machine learning for you to build an application you need data. That's the power of it, that AI can learn from data, and it actually turns out you can learn from imagery-based data. So, you can imagine an application like Aided Threat Recognition, we call it Automatic Threat Recognition, and, in this case, you need a bunch of training data. So, in this case, you need a large pool of training examples. The first step, then, is to gather that type of data, annotate it as you see here with those kinds of colored boxes, bounding boxes around the areas of interest within those images, and then annotate it, like what that object is.

In case, you see t72s btr80s, and with that, the AI over numerous iterations, especially in the context of a neural network, will adjust and start to learn those features of that particular object. Therefore, once you're done with the training, you can deploy this AI model potentially on a ground-based computer that's maybe processing satellite imagery, or it could be a robotic combat vehicle, for instance, and you are processing imagery from those sensors on that robotic combat vehicle similar to this case. And then when you present new information or new data to that model, it will actually present the detection of that particular object and cue the operator to a potential target.

You can imagine extrapolating this; as we start to become more advanced and we start pushing more and more sensors across the battle space, we're going to be facing the situation where we have unprecedented situational awareness. If we start to export AI to all of these types of systems, at the edge, it is going to give us an enhanced situation awareness. So, what do we do with that?

One, besides giving you advanced, enhanced situation awareness, you could exploit this in targeting potentially. There are lots of different targeting frameworks; you see one example, an Army Doctrine called F2T2-EA; it's fine fixed track, and I've already led into the detection stage. This fine fix track is an example of AI and how to leverage it. In this context, you would trigger an alert to an operator, keeping a human in the loop, doing this inbounds with our ethics and doing this responsibly. And part of the study was, how do we do this, a more adaptive machine team, human-machine teaming framework? And how do we speed up? So ultimately, AI is going to allow us to process many more targets much more quickly than we have been able to do in the past, and that's the big power of this.

You can imagine in large-scale combat operations, for example, there are going to be many more targets on the battlefield, and you can imagine also space-based sensors can scan a large swath of area very quickly, and you could be presented once, AI can move at machine speed, can detect thousands of targets. So how do you handle that? And this is part of the study, how do we get faster with AI? And not just necessarily cueing the operator for every single one. There might be some conditions where we could actually expedite some to a later stage in the targeting for that final verification, with maybe a targeting officer or the fired support deputy officer, to actually validate those targets before affecting them. And we have to do this faster than our adversaries because they are ultimately targeting us.

So, what challenges does this present? A few of these are listed on the slide. And the big one is data, so, when it comes down to what makes our AI better than our adversaries, there are really two things. It's our data, and it's our talent of employing it, and I'll kind of talk about the talent piece here in a moment, but first the data. And with data you're going to have to have sufficient volume to data, because you can imagine when we're employing this on the battlefield there are all kinds of condition; these tanks images show you we're going to have to fight in the winter potentially, the summer, in the fall, and desert, all kinds of different environments, and those present challenges for AI.

So, if you just trained all of your data in let's say NTC Fort Irwin, all out in the desert and collected all this imagery and you try to deploy it in Europe, it's going to be off somewhat. And that's ultimately going to produce some potential errors. So, you have factors like lighting conditions and atmospheric effects, if you're talking about satellite data, and so on. So, as you start to think about all these combinations, it gets almost intractable about the volume of data required to train a robust neural network. And this is going to really lead into how we're going to get better AI. You see those performance challenges at the bottom. So, if you have input into your neural network that doesn't resemble your training data, you're going to have the potential for errors. And what kind of errors are you talking about? They can be misclassifications; you might not even detect the target, for instance, which is a big problem.

So, more about how do we overcome this? Our argument is we've got to close this loop faster than our adversaries. AI is just like other software; you have to be able to update that software over time and really more so for machine learning applications. As I kind of alluded to, when you're going to be employing these systems, you're going to capture new data on the battlefield and you're going to see mistakes being made. We have to capture those mistakes, and we have to add them to our training pool. Then, when we can learn from those informative samples, we add them, we retrain a new model; we test and evaluate it and then we deploy it. And then next that model will perform better than the previous one. And so, this is where we're going to get our advantage, doing this faster than our adversaries, having better AI, so we're going to have to do this at the edge.

To kind of paint this picture a little bit more clearly, here is a scenario: an operational visual—you probably want to change the backdrop to more of a European theater—but you can imagine these tanks here and these UAV's employing this aided threat recognition capability at the edge, alerting operators to potential targets. As I said, maybe the adversary has camouflage, maybe they're changing some of these metamaterials, changing the look or dynamic of what those targets look like. And we're going to have to capture that data and bring it back, put it onto our tactical cloud package node and start the curation process, add it to our training pool, retrain, and update a new model to that AI.

And one thing I want to point out is, look in the upper right, where I talk about the cloud, the enterprise cloud. We're going to have to do this at the edge, but why? And we can't count on our tactical networks to push the volume of data that we need back to the enterprise, back to the states. So, we're going to have to do this at the tactical edge. Therefore, if we're doing this at the tactical edge, we can't count necessarily on contractors to do this type of work for us. We're going to have to have soldiers and all service members, in general, doing this type of work out in the field. Therefore,, we need to educate the military force.

So, this is one model that's been introduced by the Artificial Intelligence Integration Center in Pittsburgh, which is strategically located there for a number of reasons. One of them is that Carnegie Mellon University is there. I had the privilege of serving with this unit before War College, and I'm going to be serving with them when I return and graduate. So in this model, you can see the pyramid, and if you look down the triangle, the types of skill sets you see here for each one are proportional to the volume of education that's required. So starting at the top, that's where you have your senior military leaders—think of a two-day boot camp where they go there and get a crash course on AI.

Moving a little further down, you have the data analysts, data engineers, and autonomous systems engineers like myself who are going to get master's degrees and even PhDs. And they're going to be required, they're going to be really overseeing the whole process I talked about before, that development operations cycle and implementing that. But supporting them are the technicians and these are enlisted members, these are warrant officers, even officers, and in that model, they come for about six months of programming, and cloud technologies, and databases and so on, and getting the basic skills and then putting them to work where they're actually working on projects to get that practical experience. And lastly, the AI users because they're going to have to implement these systems and actually help with that feedback process, and they see mistakes, so let's capture those so we can learn from them.

And then lastly, once we build this pipeline, that's where higher education comes in, and the idea is that we export this outside of Carnegie Mellon to across the universities, and this is where you all can help with that. But once we have this pipeline built and once we we upscale the personnel to all the units, we then have got to reinforce that, we have to rehearse it, we have to practice it. That's where we get to the world-class training. We have world-class training institutions, but now once we have start to fill in AI-enabled systems, we have to go through that iteration process of updating our systems out in the field, training like we're going to fight.

Dr. Anthony Pfaff:

All right, I'll just kind of wrap up really quick and bring it back down to give you an idea of what the challenges are, particularly for these grassroots efforts. Here's the talent pool that they're drawing on right now that's organic to them. At the core level, they have seven people with technical backgrounds, but they're in space and simulations. Regarding now, taking a step back and looking at the Army in the fiscal year 2021, there are 174 advanceable schooling authorizations: 72 of them were for STEM degrees. And of course, not all of them are going to be computer science or AI or anything like that, but we can't always track those guys.

Chris is a great example; Chris's Ph.D. is in computer science, so on his record, it says computer science; it doesn't say with the concentration of AI and robotics. So, if we want to go find Chris because we really need one of those right now at the core level, we can't. He foolishly volunteered his services and raised his hand, but there's a lot to do there, and the programs that Chris was talking about are sort of intended to address some of that, but the program he was talking about with the master's degree, Carnegie Mellon, their first 20 graduates are this summer, but we'll probably need more if we want to do this at scale.

And the final thing, where this kind of ties in with industry, with the process itself, because that expertise doesn't exist in the Army; there's greater reliance on vendors and people who actually know how to build and use and manage the algorithms. And do that in the process where they haven't really been before, so that forces us to wrestle with the question, how do we better interact with industry and with academia? And we have to confront the question of how do people who are dedicated and who are very focused on that specific technology, how do we integrate them into the process? If at all. And the way I boil it down is not only do we need to find the right kind of people and put them in the right kind of places but we've also got to figure out who the right kind of people are and where the right places are. And on that triple note, we are done.

[See Appendix for corresponding PowerPoint presentations.]

Steven Weldon:

Thank you, Dr. Pfaff, thank you, Dr. Lowrance, and on time and under budget, amazing. All right, I'd like to now turn it over to Dr. Mertoguno.

Dr. J. Sukarno Mertoguno:

I looked at the symposium website, and I saw the question, so I'm going to answer those questions. So, upfront, what is the relationship, the synergy, between academia and national security? I used to be a program officer or program manager at the Office of Naval Research, where I basically built the cyber program in 2010 until I left and went to Georgia Tech in 2019, and grew the program significantly.

What I saw as the synergy, was that the military provides an interesting applications for research to focus on, and also providing funding to perform the research. You can see this from the call from DARPA, ONR, and AFOSR, and ARO. The academia can take advantage of the research calls and build cutting-edge technology. Securing research funding will allow you to grow and explore the scientific research and then develop the cool applications to fulfill what the military wants. Besides that, academia also builds a high-tech workforce and the military can provide a career opportunity in the military or DOD civilians within the labs, within the FFRDC, etc.

When I talk to these military labs or FFRDC, one thing that they can offer these highly educated people from academia is they can offer an area of jobs that are interesting, that you can't find nowhere else, not in the industry, in academia, somewhat, when you try to answer a problem that the DOD post. That's the kind of the synergy between those two.

First of all, we know that science and technology is an important factor for high capability and effectiveness for the military operation. Also, all of this high technology equipment that is built significantly contributes to the strength of the DOD, of the military, and the leading universities provided cutting-edge solutions and often help the military to develop future capability. I know for sure in the area that I work on, which is cyber security, the cutting edge of that technology is with the university, not with the labs, not with the military industry, not with the commercial industry, we know that. I saw that when I was at ONR. When I look at some of the DARPA programs, when those programs are mostly done or mostly performed by a company, the technology is somewhat not very interesting. Generally programs that is on the cutting edge, requires a lot of academic performers, a lot of professors working with the company.

And for using all of these technologies, it is important that the user of the technology, which is the military, have a high comprehension of the technology itself and know what it means. What is the assumption? What is being deployed? Asking the assumption is very critical. One example that I have seen, since we are talking about AI; around 2016 or 2017, I funded some people at the University of Washington. They attacked google perspective; this is the google tools that is used to do sentiment analysis. Behind it, there is a natural language processing implemented completely as machine learning. What happened was that it completely falls flat if you expose it with negation. So, machine learning doesn't work with negation, and you can see why, if you really think about it. Machine learning or the current AI machine learning is simply statistical machinery, and what does the statistical machinery do if you give data? It's going to look at the data, put a frequency based score in each of them and also the score among the relations and among the terms and then, in the end, it gives a final sum of scores as answer to the question/problem posed. So, if you look at a negation, say you have a sentence words as follow: a, b, c, d, 'not', e, f. You can see that 'not' is only one out of so many (7) words in the sentence. But the contribution of 'not' is actually not a summation, it's not cumulative, it's transformative because it's basically the reverse of (minus one multiplied to) whatever it is attached to. Statistics based machine learning cannot do that, as by definition statistics cannot do that.

So obviously, google has to give up; they actually said that that tool cannot be used without crowdsourcing. When I saw that, I kind of know what to do about it because I know what is behind issue, the statistics. Statistics can only deals with accumulation/ frequency of occurrence. So, I kind of have a potential solution in my mind of how to deal with it. I was at ONR, I don't have anybody to work on it, and that's too simple of a problem for me to give it to a professor to explore it, so I just leave it. Recently, I'm at Georgia Tech now, I have students, I have my postdoc joining. There was two weeks delay due to paperwork doesn't get done, before he can officially join. So I give that problem & my idea to him, and within that two weeks he comes back and officially joined Georgia Tech, he got it done.

Whatever idea I gave him was correct. This is basically what do you do. So, okay you have a bunch of sentences, you kind of have to know where the word 'not' is attached to; that's the first thing. So, once you do that, you remove the 'not' and then you send it back to the google perspective, when you get the scores back, you revers the scores (multiply with minus one), all the scores of the phrase where the word 'not' is attached to, and sum all of the scores into final score. But these are a problem that by definition machine learning cannot solve no matter how much you train. Why? So, at first, you're learning you can put a 'not' there, but the word 'not' can occur naturally in so many different locations in the sentence. Now, if you want to include a possible occurrences of the word 'not', your data set is going to blow up. Also, if you expand your data set with the potential occurrences of 'not', you also have to correctly have to label it; that's also another problem.

So, you see there that one needs to be careful, one needs to think about the problem. I think understanding the problem is the key here. So I think that what we need to teach people is actually analyzing and understanding the problem and also obviously, people need to have a lot a scientific background so that when they understand the problem, they know how to solve it right. It requires both. So, then to the next point is that exposure to scientific research can be very significant for the military people. When I was at ONR, we used to send some of the STEM money to Annapolis (Naval Academy), we partnered them with the UMBC (University of Maryland Baltimore County) because of their close proximity, so now the Naval Academy professors can do research. Since Annapolis is undergrad school, so it's hard for the professors there to do research. Now ONR sponsored partnership with UMBC, they can do research; they have research collaboration with UMBC, they have access to graduate students, but they also It provides some of the cadets to research, which is really nice. It expose them early on how to actually do scientific research. I think that will be very useful for their future career when they become officers. That's what is important, I think.

So what is the future of the relationship? The future of the relationship between the military and academia has been basically established and cemented by congress with the ONR being established in 1946, DARPA in 1958, and these are the DOD part that deals with trying to bridge innovation between the military and the academic. And the funding and the size goes up and down, based on whatever the mood and the politics within the Pentagon is, but it will not go away. It will always be there I believe. And NSF also helps in the building of scientific foundation. So, I believe that the strong collaboration between the academic and military is mutually beneficial and strengthens both sides in the growth of technology and the human capital toward national security.

So what is the problem with the higher education and national security? Well, the biggest problem is that a small percentage of US citizens are pursuing a higher degree. Part of it is that economic push, after college, one have to choose whether to get a job or go back to school (for graduate degree). It's kind of difficult decision to make. There are programs that address this issue but I think part of it is that to be able to work in technological part of the DOD, normally needs clearance. So, getting people who can be cleared into the pipeline will be great. It's hard to achieve, so that's the thing that I can see as a problem.

So what do other countries do? I believe that the US still is probably one of the countries that provides the best synergy between the academic and the DOD. And in the more direct way, with a call for certain research, etc., thats what DARPA and ONR and AFOSR and ARO can do right. That's kind of a direct interaction to move the science for the military. Other countries do that too; the UK and Australia do that on a much smaller budget. The EU, from what I understand, are more of trying to do all of these funded research toward the industry rather than the military, that's what I saw. And China I don't know, so the way the country is, it's probably a little easier for them to push some of this, but I have no data. I think that's all.

[See Appendix for corresponding PowerPoint presentations.]

Steven Weldon:
Thank you, Dr. Mertoguno. Director Toler, you're up.

Colonel Eric Toler:
I was going to start off this pitch with, I really think that artificial intelligence can solve all these hard problems we just talked about today. But then last night when I was coming in, I put in google maps how to get to this place, and it tried to route me through the cemetery, and I'm like, okay, maybe that's not the case. And certainly, after the presentations we've heard today we know that's not the case. So, I'm going to do a recap really quick on what we're here to talk about. I just highlighted some things I think are important. So, we talk with the resurgence of artificial intelligence; I guess that means there was a surge to begin with. Now we're resurging. We must look at ways of employing the technology better than our adversaries. It takes skilled people to update AI enabled systems; they have to adapt their AI systems during deployment. Lord help that's hard. We need service members who continuously build, test, and deploy new updates to their AI enabled platforms. I've never seen that. Thus, the way forward, even if it's only in an interim stage, is for DOD to partner with academia and industry.

Correct me if I'm wrong, but I heard General Brown this morning say that the military and higher education relationship is critical to national security; did I get that right? Yeah, does anybody disagree with that? Okay, that's pretty powerful. That's critical to national security. I will tell you I've been out of the military for three and a half years now, and I believe that's true much more now than it was three and a half years ago. I also heard him say DOD and government can't move fast enough to keep up with technology and our adversaries. We don't have a very good cultural understanding of war in the US. I agree with that; I disagree that war has changed or is changing. I think our understanding of war has changed and the way we use tools may have changed, but there's still a lot of pain and suffering that goes along with war, and for those who read military history and Sun Tzu, you know how to win without fighting. I think we're seeing that unfold as we speak; bureaucracies are designed to say no.

I had the pleasure to work in the Pentagon for three years from 2006 to 2009. That was an important time for cyber because that's when we defined it; we haven't defined AI yet, but the Secretary of Defense defined cyberspace and cyberspace operations (as military terms), and I will tell you as we have leapt forward 15 years, we haven't fundamentally changed the way we recruit, train, and retain talent, as well as the way we do capabilities development, or we fight as a nation in cyber. Can anybody tell me who's in charge of defending the United States against a cyber-attack? No? Okay, so back to the topic. As an intelligence officer for 24 years – I was an artillery officer for four – we've always had a fascination with AI machine learning or, better said, data analytics or data correlation. What we really want is to put all these sensors out there and use these smart sensors and this data correlation to tell us exactly what the enemy is going to do and when they're going to do it, etc. And that's never worked.

I'll go back to a system that we fielded over 30 years ago. The All-Source Analysis System (ASAS), which was pretty cutting edge at the time, had the ability to take in all these reports from intelligence in a certain format, and they could correlate that data – they could take location, they could take unit names, and correlate that into a single unit or icon on a map; I mean, it was really cool. The only problem was it didn't work in the field they way it worked in exercises. And no one really understood the technology behind it. But you could send your soldiers to a five-week course – I think they call it the ASAS Master Gunner Course because we have to call everything by military terms. However, the course focused on how to modify and update the system and make it so that the data flows worked, and if you had an issue, you could fix it. That worked extremely well, so in exercises we actually made the system work, but then we deployed to Bosnia. It wasn't made to work in Bosnia (in a peace enforcement operation), so we didn't use it. So, we said we need to make this change; they (the contractor) said that's great, write it down, and we'll do a change request, and in two years, we will change it for you, and we'll charge you another 100 million dollars (embellishing).

So I would argue that, although that technology was pretty cutting edge back then, that's not a good way to do business. I had an epiphany in 2004, however, when I deployed to Iraq – in this case for an Intelligence agency, and we were putting out sensors that were just programmable receivers with a server rack and antenna suite; they just collected RF energy. We had a technology team on the backend (in the States) who were computer engineers, so while we were forward, they were writing the code that created the graphical user interface and all the data flow algorithms that we needed. And I was like, man, now this is the way we need to do business. And we would tell the contractors, hey, we'd like for it to do this, we need it to do that, and in a matter of minutes or hours, they were able to make those updates, and we had this very nice collection system that was tailored to the need at the time. Also, at the same time, Army units were told they couldn't deploy with their program of record equipment for SIGINT collection because it was completely useless in that environment. So, I just use that as an example, and I apologize I don't have any slides, so I'm just going to tell you stories, I hope that's okay.

Going back to my artillery days, I was actually a Division Artillery S2 at the time, when we fielded the paladin system. You're probably familiar with the (older) cannon system we had, where everything was very manual. We had to input meteorological data and muzzle velocity data, which essentially equated to the different lots of powder because every powder kind of burned at a different rate, so it shot further or shorter depending on the chemical makeup of the powder, powder temperature mattered, etc. So this new paladin had a GPS in it; you pull up, it computes all your firing data, registers your muzzle velocities when it shoots, so in a matter of seconds you'll be able to fire on target. The only problem was when we fielded the systems, we couldn't hit the target, and guess what, nobody could figure out why, because we had stopped teaching manual gunnery and how all those components worked. So, I'm going to get to a point on that in a second. And then same thing with cyber security, so we have lots of algorithms that do monitoring and can detect malware as it sees it, action that malware as it sees it, but if you don't have the fundamental understanding of how malware works and how networks work and how data transfers over networks, you don't understand how to operate the capability.

And the last story I'm going to tell you is about the most incredible intelligence capability I ever witnessed, which was at an organization that I was very blessed to lead at Fort Gordon. On our watch floor, we had software developers and engineers who worked on this platform that operated in a state that we called a perpetual beta environment. So, I've never seen another DOD system work in that manner . We also had a researcher, a graduate from Georgia Tech, but she worked in the organization's Research Directorate, and she was able to pull technology that we were looking at 10 years down the road into the current fight because she came in and asked what we needed. Well, we need this, we need this voice recognition software, we need this signal recognition software, to be able to pick out, you know, the one Russian speaker amongst a million Arab speakers, which just happened to be the number three guy in ISIS who was a Chechen. So that technology and the way we used it was phenomenal.

So, there are four things that I'm recommending that we do, and I think we've touched on some of these. One is we absolutely have to educate our operational force on technology. You don't need to be able to write algorithms, but you need to understand the foundation of the technology. You need to understand the data and how to normalize data, and what data means, and what data you need in order to achieve solutions and outcomes.

Number two is to organize into teams with dedicated technical support just like I talked about when we had two NRO engineers who were in direct support of our teams in Iraq. This is important; we talked about how we don't reward cultural expertise; but we don't reward any expertise in DOD right now – but having that is critical. And just like we provided reach back intelligence support, which again was very effective, we also need that technology support to be able to make those changes that you talked about.

And then, same thing – cultivate a developmental operations culture. That means, when we get into academic research, they need to be informed by whoever they're supporting. And when we've done that well, it works very well. But I've also seen, in far too many cases, where research goes off on a tangent, both in government and academia, and the outcomes are not optimal.

And then the fourth thing I'll say we need is to incentivize the expertise. We talked about how critical it is in government, but we are losing our intellectual capital in the US right now. You know, a way (to gain intellectual capital) is to give our PhD graduates green cards, but we need to figure out how to capture domestic talent, as well. I just did a quick search on computer scientists and electrical engineering, just picking those two topics in the US, and 81% of our graduate students are international students. I would argue that's far too skewed. I think 10 years ago it was about 50-50. That's where it probably needs to be because then you've got the best of both worlds, but right now if you get a computer science degree, you can go get a really good job in the private sector, so how do you incentivize students to pursue graduate programs?

And I'll tell you, time is of the essence, so as an intelligence officer who did multiple cyber operations tours with Army Cyber, US Cyber, and then 10 years with NSA, I watched China go from a near peer to a peer power, and they're on a glide path to overtake us (in the next 10 years). You argue with me on that — but I've probably seen a little bit more data than y'all have — and I don't like to lose. And so, unless we can fundamentally change the way we're doing business, we are going to lose. The advantage we have is our innovation culture as a country; that's what a free society does. You know if you're a communist regime or other dictatorial regime, there is really no separation between government, academia, and private industries; it's all kind of one thing, but it's hard to innovate. And so, I think we have that advantage.

And then last thing that I think our universities need to do is much more outreach. I think it was touched on earlier, on educating our population, and assisting with K-12 (education). We can't just sit back and wait for students to come to our wonderful higher education institutions, because our lower education is terrible in many cases, and that's just my experience. We don't teach problem solving and critical thinking skills in some of our secondary education systems any more. So those are just some of my points that I've seen in my three and a half years at the Georgia Cyber Center. I mean, we're really focused on cyber operations and security, but I think as technology and engineering and STEM in general, these are some issues we have to tackle. Thank you.

Steven Weldon:

Thank you, Director, Toler. Dr. Parlier, over to you.

Dr. Greg Parlier:

I have about 10 minutes of narrative script hopefully synchronized with about half a dozen or a dozen charts, and I'll default to a lecture mode here, which is totally inconsistent with the Socratic method of teaching that was imposed upon us when we were young, but try to get a lot of information out here quickly. So, you'll see that it's just a variation on a persistent theme here;

it's all about innovation, how to generate innovation in an era of emerging technologies and then how to accelerate innovation in the context of the military and its interface with academia, the corporate world, supporting the military.

First, some history regarding the role and contributions of military operations research – what we refer to as "OR" if you haven't heard the phrase before. Early during World War II, a new multi-disciplinary approach for solving complex military problems was pioneered by the British; Operational Research, OR, combined civilian scientific talent with Royal Air Force military staffs, initially to support fighter command's urgent preparations for what would soon become the existential battle of Britain or rapidly gain credibility within the RAF and quickly spread to support the US Army both ground and air forces as well as British and US naval forces. There's much we can learn, including enduring principles from these early World War II years when OR was conceived to integrate new technologies, such as radio detection and ranging —which we simply refer to today by its acronym RADAR—into combat systems, operational command and control for the RAF, and strategic defense during the Battle of Britain. The idea for rapid learning using a system of teams defined and differentiated OR at its inception with expertise across a wide range of scientific and engineering disciplines. Using empirical evidence from ongoing military operations in conjunction with creative mathematical models for rapid learning, OR represented a technological advancement unique in the history of military decision making. Working closely with, trusted by, and responsibly advising high-level commanders and government leaders, all while operating under extraordinary pressures, was a hallmark of OR from its very beginning.

Now, nearly 85 years later following two decades of conflict, we are experiencing another post-war transformational challenge on a unique cusp of history. Successfully integrating emerging technologies into weapon systems, operational concepts, and strategic plans as a central challenge confronting military innovation. Today as then with the example of radar we're confronted with comparable challenges to integrate these emerging technologies into combat platforms systems and strategies, among them robotics and autonomous, systems, AI and machine learning, micro-electromechanical systems and nano-technology, and, of course, hypersonic and directed energy. At a time when military operations research, both the practice and the community, appears to be at a crossroads, the trajectory of this unique problem-solving discipline must be realigned to current and foreseeable challenges.

Military organizations, especially successful ones, are renowned for their strong cultures, yet the long history of military innovation reveals those cultures can also become impediments to organizational adaptation, when failure looms. Although scientific advancements continue to amaze us, we must better understand how technology, management, and policy interact in our complex socio-technical enterprise systems. Management innovation often lags technology advances, yet is essential to fully capitalize on rapidly growing big data opportunities. A strategic analytics framework aligns the ends, ways, and means strategy paradigm with corresponding prescriptive, predictive, and descriptive analytic domains, focusing on the ultimate purpose for which an organization exists, descriptive analyses, segment problems, diagnose structural disorders, and identify enabling remedies and potential catalysts for innovation. Next, a system-wide integrating perspective synthesis addresses the attainment of enterprise goals and objectives, desired ends, if you will, using prescriptive analytics. Finally, the design and evaluation phase provide comprehensive road maps using predictive analytics to create analytical architectures or the ways to guide transformation.

Among the enabling disciplines, capabilities, and methods for strategic analytics are decision support systems, "engineering systems" - to be differentiated from traditional systems engineering, dynamic strategic planning, and engines for innovation to enable rapid experimentation, generate insight, climb steep learning curves, and develop strategies around new concepts and technologies. Although so-called IT solutions have ubiquitous appeal and enormous investment levels, we need to include analytical architecture for enterprise challenges. Without the integrative power of OR to focus process re-engineering on desired outcomes, this obsession on IT can result in growing complexity and exceed the interpretive capacities of organizations. Ultimately, it is management innovation that will enable better decisions from the growing amounts of information and improve situational awareness made available by advances in IT.

Many of our systems seem fragile and vulnerable increasingly subject to catastrophic failure due to age and decay, human error, whether negligent or deliberate, or what is known as tight coupling in complex systems. And while traditional engineering methods optimize performance based upon design specifications within assumed operating environments, our own experience and history reveal that these systems and their use change over time, often in untested unanticipated ways. Thus, a capacity for adaptation must be built in to create a resilient system that can adjust as needed. Just as nanotechnology is increasing our understanding of very small-scale structures, the evolving discipline of engineering systems is expanding our macroscopic understanding of very large-scale enterprise systems, defined by their technical managerial and social complexity. This new and evolving approach represents a new paradigm in systems design by shifting from the traditional focus on fixed specifications, or what we call military "requirements", toward the active management of uncertainty in the implementation of socio-technical systems. Most system design methods generate a precise optimized solution based upon a set of very specific conditions assumptions and forecasts. However, these methods are rarely valid over longer planning horizons as strategic designs for technological systems. In contrast, dynamic strategic planning instead presumes forecasts to be inherently inaccurate. DSP therefore generates flexibility by building in adaptability to changing circumstances that inevitably prevail. This built-in flexibility creates additional value for the system which, in many cases, can be quantified. So, this capacity for adaptation enables a resilient enterprise that can then adjust gracefully as needed rather than suffer slow-motion or catastrophic failure.

How, then, can innovation be better understood and accelerated in a controlled way to minimize the debilitating effects of disruption? While institutional adaptation requires a culture of innovation, inertia remains a powerful force within bureaucratic organizations. An engine for innovation or EFI is a virtual test bed needed to provide a synthetic non-intrusive environment for experimentation and evaluation of creative ideas and concepts. This synthetic environment or micro world transforms theoretical knowledge into practical applications by catalyzing innovation often found in the seams between disciplines, technologies, and institutions. EFI's also minimize the debilitating effects of disruption. They provide sources for socio-technical innovation to expand organizational capacity for social ingenuity, identifying implementation issues before they're adopted as policy and institutionalized across the enterprise. Thus an EFI generates technological and managerial initiatives consistent with the organization's vision, incubates and rigorously analyzes them within a non-intrusive test bed, then rapidly transitions into actual practice those selected as most promising.

The functional design for an EFI includes three organizational components that comprise core competencies or mission essential tasks. These three components perform the following: they encourage and capture a wide variety of inventions, incubate those great ideas and concepts within virtual organizations to test, evaluate, refine, and assess their potential costs system effects and contributions in a non-intrusive manner, and then transition those most promising into actual commercial or governmental practice. The purpose of this deliberative cyclical discovery process is to sustain continuous improvement through experimentation, prototyping, field testing, and rigorous analysis. Innovation engines accelerate organizational learning while encouraging both technological and social ingenuity as foundations upon which national power can be generated and sustained in the future.

We must remember that a major precept of any learning organization is an ability to genuinely learn lessons from the past and then actually apply them rather than merely observe them. Such a retrospective can reinforce rather than delay innovation by discovering enduring principles that should be resurrected and applied. One possible framework to rigorously assess the current state of operations research is to apply enduring principles derived from the early experience of OR during World War II, among them, capacity, capability, organization, utilization, and contribution. So I put this particular chart in just to, first of all thank AUSA—they're one of the sponsors for this particular symposium—but also to highlight that one person's perspective on those five different methods of evaluating OR or assessing it today within the military is available for you, and I got a couple copies of that, can provide you the link if you're interested.

In recent years the application of strategic analytics to several Army enterprise challenges has shown that engines for innovation can be valuable organizational mechanisms for successfully pursuing transformational strategies. Central to these endeavors was the extensive application of OR, data sciences, and management innovation for improved performance. Although their fundamental natures were vastly different, ranging from defense resource planning to sustaining our all-volunteer force which we almost lost in the late 90s and then more recently to transforming our supply chains particularly our materiel or sustainment supply chains. They all required an ability to organize, manage, lead, and develop highly talented multi-disciplinary teams. These new concepts and methods for strategic analytics should now be extended and applied more broadly across many other national security challenges as well.

In conclusion, operations research can provide a crucial, indeed unique, source of American power. We should renew and restore OR as a core competency for military innovation, defense planning, and operations analysis. Strategic analytics can be used to illuminate better ways ahead for defense modernization and should be adopted to encourage imagination, confront conventional wisdom, and better reconcile ends with means in the face of major national defense and international security challenges.

[See Appendix for corresponding PowerPoint presentations.]

7

MILITARY, ACADEMIA, AND HISTORY: A VITAL 21ST CENTURY TRINITY

Major General Mick Ryan

As presented at the 2022 Strategic and Security Studies Symposium
Hosted by the Institute for Leadership and Strategic Studies
University of North Georgia

Dr. Keith Antonia:

At this time, I'd like to introduce Dr. Ken Gleiman, who is a retired army colonel and also an author of a book called *Operational Art and The Clash Of Cultures Post Mortem On Special Operations As A Seventh War Fighting Function* which he published in 2012. He is currently president of the Army Strategist Association and vice president of Inspirata Consulting, a company that evaluates programs and strategies for the United States government. He also teaches at Georgetown University. He served as a special forces officer and an army strategist for 27 years. He received his Ph.D. from Kansas State University. Through the Army's Good Pastor Scholars program, he is also a graduate of the School of Advanced Military Studies, as well as Georgetown University School of Public Policy and the Army's Art of War program. He served in multiple combat zones and at the Pentagon and combatant command level assignments.

Dr. Ken Gleiman:

My thanks to everyone here at the University of North Georgia for putting on this symposium. I have the pleasure and honor of introducing Major General Mick Ryan of the Australian Army.

I like to say that Mick is probably the most famous Australian general officer in the United States, and I think that's for several reasons. Of course, he spent a lot of time here in the United States as a graduate of some of our civilian institutions and our military institutions, but mostly he's been a passionate advocate of professional education and lifelong learning. In 2021, he was an adjunct scholar at the Modern War Institute at West Point, and he's also been a huge supporter and contributor to many of the forums and blogs that a lot of us belong to, especially a lot of us in the army strategist community. Most recently he's provided some of the smartest commentaries on the Russia-Ukraine war that I've seen, so if you get the chance to read some of that afterward, I highly recommend it. He's a strategist, author, and speaker, and has a distinctive mixture of experience and skill. He grew up in a small mining town in Central Queensland before attending the ADF Academy and the Royal Military College at Duntroon. He has over 30 years of working in dynamic groups focused on overcoming adverse circumstances to solve complex institutional problems. So, whether it's been in a small town supplying clean drinking water or leading construction efforts in southern Afghanistan, or managing institutional change management efforts for the ADF, he's really the kind of individual who's most comfortable when he's part of a professional, diverse team focusing on really challenging problems. His early career consisted of a range of appointments in operations, training, and organizational development, and he served internationally in East Timor, Indonesia, in the United States, as I've already mentioned, and, in both Iraq and Afghanistan, as well as some various exchange postings with the United States Marine Corps. In more senior appointments, Mick led the development and execution of large-scale organizational reform programs for the ADF, including defense-wide education and training from 2018 to 2021. He also represented Australia on a secondment in the

Pentagon where he worked on strategy for the Chairman of the Joint Chiefs and for the Obama Administration.

Mick likes to say that his first love for over 35 years has been investing in people, and he definitely demonstrates that all the time. In this technological era, many forget that it's people upon which companies and institutions are founded, and to that end, Mick has a deep and abiding commitment to leading and personally investing in people, being an exemplar for their continuous learning, and being their most passionate advocate. He, as I said before, is deeply committed to learning as a lifelong undertaking. He seeks to set the example for those he leads in this regard while being a recognized global expert and influencer in professional military education, strategic military planning, institutional reform, and adaptation and leadership. He was made a member of the order of Australia for his leadership of Australia's first reconstruction task force in Afghanistan. He holds master's degrees in operational studies, public policy, and military studies from the United States Marine Corps and from Johns Hopkins University. He's married and has two daughters, and just in case he's too humble to plug his book, I'm going to plug it right now because I've just read it. His new book is called *War Transformed*; it was published just this February by the USNI Books. So Mick, welcome!

Major General (Retired) Mick Ryan:

Thank you for that wonderful introduction. It's very humbling. It's lovely to be with you, I do apologize for not being there in person, but I just wasn't able to get a plane there this week. I am keen to talk to you all. The past six weeks have reacquainted us all with large-scale state-owned state warfare; the Russian invasion of Ukraine and the state Ukrainian defense of their country has provided a broad range of strategic, operational, and tactical lessons for us to study for months and years to come, and it's a story that's almost too good to be true. The massive army of an authoritarian tyrant meets its fate at the hands of a small democratic nation led by a charismatic former comedian. I mean no Hollywood scriptwriter would dare write such a story. And the reason I raise this is that it allows more people than ever to observe and learn lessons from war and to do it in almost real-time.

In some respects, it provides us with a similar opportunity to what the 1973 Arab-Israeli War— fought in the immediate aftermath of the Vietnam War—provided the United States army, providing the lessons, the wake-up call, and the impetus to move away from counterinsurgency back to complex combined arms operations. The conflict provided a range of very useful insights into a modern war about airline integration, anti-armor missiles, electronic warfare, the increased lethality of infantry, and other capabilities that would become core aspects of the US Army's modernization post-Vietnam. The lessons learned from that conflict were an important input for one of the most successful military transformation activities of the 20th century. The classic 1983 military review article by General Donn Starry *To Change an Army*, which I'm sure many of you have read, describes this reformation of the United States Army much better than I ever could, but as we observe the Ukraine war from afar, Starry's final statement in that article is worthy of repeating. He wrote that "the need to change will ever be with us. The intellectual search and exchange of ideas and the conceptual maturation must continue and be ever in motion."

We should therefore view our lessons from the current Russo-Ukraine war as part of an exchange of ideas and conceptual ferment that remains in motion. Like all wars, this features some old things and some new things. Important continuities we have seen in the war include the centrality of land forces in fighting and winning wars, the need for effective combined arms teams and airline integration, and the requirement for an effective logistics system at every level, but perhaps the most important lesson, one that is as old as war, is that good leadership is the most important aspect of successful military organizations. When soldiers are given purpose and led by honest, humble, and courageous leaders, they can and will achieve anything, and as we've seen in Ukraine, the Russians have deployed one of history's worst-led armies. Ukrainians gave massive disparities in size equipment and resources between them, and the Russians fought one of the best left defensive campaigns in modern history.

What does this all have to do with what I'm supposed to talk about with you today? Well, I believe that the Ukrainians are a wonderful and a very contemporary example of my core topic: that integrating national capabilities with a country's military instrument can provide a significant capability edge and decisively alter the outcome of a conflict. The Ukrainians have shown us what it is to mobilize a nation in the 21st century in a total effort to defend their nation. They've integrated military diplomatic information, economic, and other efforts into a single unified approach led by their president.

I'd like to focus on one aspect of the integration of military and non-military national assets today. The area I will focus on is the military industry-academic team and how integrating these sectors and their efforts better provides a nation with a greater edge in both strategic competition and the walls we are likely to face in the 21st century. In my talk, I'd like to start with a quick examination of trends that are having an impact on 21st-century competition and conflict, and where these are in evidence in Ukraine, I'll call them out. I'll follow this with an examination of the imperatives for closer integration of military industry and academic efforts. This will be followed by an exploration of the steps my own nation has taken in the past decade to achieve this alignment of national assets for national security outcome. But first, what is 21st-century competition war going to look like? It's important for us to understand this because it's the context in which the military industry-academia team will function beyond the lessons from Ukraine.

Military, Academia, and History: A Vital 21st Century Trinity

We have been given many indications about how the strategic environment is changing and what this means for military institutions. Even before Covid, we'd already lived in a world where superpower competition had returned; the last two years have only supercharged that competition. They've also made very clear to us the aspirations of China as a global, technologically advanced, and very wealthy autocracy. Its coercive behaviors, secrecy around Covid, and its obsession with suppressing any form of dissent at home and abroad, all bode ill for the coming years and decades. At the same time, it is the richest, the most technologically advanced competitor that the United States has ever faced, and this competition is set against the background of massive changes in technology and society. Rapidly advancing technologies in the 21st century in their military applications can at times be overwhelming. There are some, such as AI, autonomous systems, quant technology, and space technologies among them, that herald changes in the character of war. The confluence of these new technologies, increased strategic competition, and the development of new Chinese, Russian, and Western thinking about war has resulted in what I believe are seven important trends in 21st-century conflict.

I'd like to briefly describe these as they are an important context for the topic of this talk. These seven trends are as follows: first, the battle of signatures where military organizations must minimize their tactical to strategic signatures, use recorded signatures to deceive, and be able to detect and exploit adversary signatures across all the domains in which humans compete and fight.

Second, new forms of, new approaches to mass manufacturing using 3D printing, and the ubiquity of autonomous systems across the land, sea, air, and space domains that herald a new era of war and competition. Successful 21st century military organizations must build forces with the right balance of expensive platforms and cheaper, smaller, autonomous systems that'll be more able and more adaptable to different missions and be more widely available.

Third, more integrated thinking in action. Over the past century, the domains in which humans compete and fight have expanded. Air space and cyberspace have joined the age-old domains of land and sea conflict, and, unlike the counter-insurgencies of the past two decades, our military institutions in the future must be able to operate in all domains concurrently and integrate into a broader national strategy. We have seen the Ukrainians excel at this integrated national approach to defending their land. We've also seen the Russians totally fail at every level of integration, from combined arms team to unified command and control.

Machine integration is an important trend where robotic systems, big data, high-performance computing, and algorithms will be absorbed into military organizations in larger numbers to augment human physical and cognitive capabilities to generate greater mass—more lethal deterrent capabilities, more rapid decision making, and more effective integration. This has not been observed to the degree that many of us expected. There has been some use of aerial platforms for surveillance and attack but no use of swarms or uncrewed ground combat systems.

The next trend is the evolving fight for influence. War has always been an intricate balance of physical and moral forces. Disruptive 21st centuries' have not only enhanced the lethality of military forces at greater distances but they also now provide the technological means to target and influence various populations in a way that has never been possible before. The Ukrainians have given us a master class in their global influence campaign. If it was a game of football, the Russians might well have not turned up for the game.

The second to last trend is greater sovereign resilience. While de-globalization and re-shoring and manufacturing have been underway for some years, the Covid 19 pandemic brought to the fore the requirement for greater national resilience in supply chains for critical materials and manufacturers. Future military organizations must also mobilize people for large military and national challenges while developing secure sources of supply within national and alliance frameworks to ensure that supply chains cannot be a source of coercion by strategic competitors or potential adversaries.

And the final trend for this century is that we need to change how we understand and use time. It is an underappreciated resource, especially government service. The speed of planning, decision making, and action is increasing due to hypersonic weapons, faster media cycles impacting political decisions, and the potential for AI to speed up decision-making at many levels. Concurrently, we're facing a long-term period of cold but connected competition and conflict with China. I propose that democracies specialize in 24-hour cycles and three-to-four-year election cycles, but they're weak in dealing with microseconds and decades. This has to change.

For 21st-century military institutions, building sources of advantage will include new technology, but it is not a complete solution to the many national security and military changes of the coming decades. People and the new ideas and institutions they produce are at the heart of my examination of building military advantages. The relationship between military, academia, and industry is not new and has been historically important in most nations, but with more forethought and trust between each, this relationship can be superpowered in the 21st century to help us tackle the growing techno-authoritarian nations.

What might this integrated military industry-academia team look like? This idea of closer collaboration is hardly a new one. We only need to look at our history, especially the second world war, to see how such collaboration resulted in significant strategic and technological breakthroughs. This included not only the development of the atomic bomb but also the arrangements for immediate post-war occupation concepts, including their civil and military aspects for Germany and Japan. A more contemporary model in China uses a strategy of civil-military fusion. The Chinese have implemented a whole nation approach to innovation to create and leverage synergies between defense and commercial developments. I propose that US and western nations more generally will need to embrace a similar approach if they're to compete with Chinese innovation and rapid adaptation absorption of different technologies. This approach will facilitate a deeper appreciation of when and how national security thinking might need to evolve as well as the adaptation required for military organizations, warfighting concepts, and the development of their people.

I think they're four important elements of what a western version of this 21st-century partnership might look like. These are strategic designs for such collaboration, the mechanisms for enhanced engagement and collaboration, better integration of outsiders into a military learning continuum, and a culture that encourages learning, experimentation, and failure. I'd like to now briefly explore each of these. Firstly, strategic design. Before we jump into a more integrated relationship, we should decide what its strategic objective is; therefore, some kind of strategic design for enhanced military academia-industry relationship is needed, and a key part of this design will be a vision of what the military institutions seek to achieve in this relationship. Visions are really interesting things. One of the most insightful examinations of institutional vision is a 1990 RAND report that explored the transition of military institutions at the end of the cold war. It described the essence of vision in a military organization as the sense of identity and purpose found in an organization that provides the members with more than what they can find in corporate strategies or long-range plans.

A vision provides the essential intellectual foundations for interpreting the past, deciding what to do in the present, and facing the future. Therefore, we should start with a strategic design with an integral vision as a basis for investing in a reinvigorated intellectual development across the military academy and industrial enterprise. It should lay down key markers in why each is an element of the partnership and what each offers and what is the greater contribution this arrangement provides to the overall defense and national security outcomes. I propose that initial areas for collaboration include new defense strategies, new ways of thinking about absorbing advanced technologies, new and evolved concepts of operations, and new ways of attracting training and educating our people. The strategic environment, as we've seen in the last two years, evolves quickly, however. Therefore, this strategic vision and design must also contain priorities that are revisited regularly in periodic reviews to ensure that the balance of investment in the different elements of the three-way partnership will continue to generate the best return on investment for our nations.

The second part of this design is strategic engagement and deeper collaboration with civilian universities, and the intellectual capacity of residents in the industry, such as think tanks and R&D organizations, is critical. In these civilian institutions reside often hundreds of years of learning across the humanities and sciences. It is a knowledge resource that cannot be replicated in the vast majority of military institutions. These non-military entities represent a resource that can provide intellectual rigor to further hone the intellectual capacity of military organizations. They also provide viewpoints on national security that might differ from officially sanctioned policy, forcing the military to more carefully analyze contemporary national security policy and warfighting concepts.

There are also other aspects that make better collaboration with academia and industry attractive for the military. First, the complex integration of different and rapidly evolving technologies and ideas places a premium on higher-order cognitive skills. Academic and think tank institutions are well placed to develop and test the critical thinking and complex problem-solving skills that are required. They might also provide other sources of the continuous learning required for future officers and NCOs to remain contemporary.

And, finally, partnering with academia and industry should provide important sources of innovation for military institutions. This might cover a broad range of military technologies, concepts, and policies and could permit a better appreciation of when and how adaptation to military training and education is to take place and the levels of technological literacy required for military leaders.

The third area of this design is the integration of outsiders into military learning continua. Military institutions are good at running training and education for their people, and sometimes we even allow other government agencies and foreigners to participate. I've been a beneficiary of this in the United States myself on many occasions. This has to expand in an evolved military academic-industry collaboration. This is particularly the case with professional military education, but there's also a range of training courses where having a wider variety of students and instructors might benefit all participants. And there are two particular areas where we might make changes quickly and gain a good return on investment. First, the better use of military educational institutions such as think tanks. Military organizations should not see their professional education institutions as pure learning for individuals. High-caliber military and civilian personnel are selected to attend courses at Staff College and Moore College. Given their talent, the large range of diverse large networks of diverse officers, and their access to high-quality academic advisors, is there not a greater role for these programs in thinking about institutional responses to the strategic challenges that are disrupting national security establishments?

The other area that might make improvements to military education is to invite industry and academia into our elite programs. Institutional education programs have to balance their focus between two competing priorities. This is between honing excellence and the very brightest of military officers whilst also catering to the developmental needs of the largest waves of very good officers who largely will comprise the staff officers and mid-level managers and leaders of a military organization. Some institutions, such as the United States military services, have sought to address this through the creation of second-year elite programs, such as the US Marine Corps School of Advanced Warfighting, US Army School of Advanced Military Studies, and the Australian War College's Art of War Program. The brightest students have also been permitted to undertake year-long studies at elite universities. These programs are focused primarily on producing improved operational and institutional planners. They might also in the future have a greater focus on building better strategists and strategic leaders. This will build a smarter military leadership that will make these programs more attractive to outside agencies because just as we might use our educational institutions as think tanks with more academic and industry participation so also we could improve our elite programs with external participation, and this might be reversed to ensure more of our best thinkers participate in external programs.

Our final part of this design is building a learning culture for this military academic-industry partnership. This improved collaboration will require a culture that defines acceptable failure and allows the entire enterprise to learn from these failures. Culture is an important consideration for this relationship because each participant will bring their own culture with them to the partnership. This culture will influence military, academic, and industry organizations' success and failure in all their activities. So, for us, cultural factors determine the professionalism and discipline of individuals and teams, and these factors drive battlefield and broader military effectiveness. We, like the other two partners, must therefore have an appreciation of the culture of each partner in this relationship, and it's a necessary precursor to success. Senior leaders will need to be advocates for a collective learning culture. They must also implement a range of different incentives: promotion pathways, talent management programs, mentoring, and rewards available to participants to nurture such a culture. Wick Murray notes in the military innovation inter-war period that one can foster a military culture where those promoted to the highest ranks possess the imagination and intellectual framework to support innovation. An improved collaborative relationship between military, academia, and industry will require military institutions to carefully explore options for participants, and the incentive structure must be just right for such a relationship to succeed.

Now I come to the hard part and the final bit of my presentation. I'm going to assess my own country's efforts in achieving these four elements of better collaboration between industry, academia, and the military, and as you'll see, there are some areas of differences between Australian and American approaches, and I'll be the first to admit we have not solved every challenge. Firstly, the strategic design; well like many institutions, we don't suffer from a lack of different institutional strategies. We have them to cover every institutional possible; some of them at times are even useful. A key strategic guidance document for defense strategic update 2020 does include a useful section on the defense industry, but it's largely about building things, not thinking things up. It's supported by a series of industry plans for things such as shipbuilding. We also possess a defense learning strategy which was endorsed in 2019 by a chief of defense, but there is no overarching plan that might coordinate military interaction with industry and academia, so this would be a gap we'd need to fill.

The second part of my design, the mechanisms for enhanced engagement collaboration. I think there's been news on this front in Australia. We've recently signed new contracts for the provision of academic services to our academy and for our War College at Western Creek; both of these have created an opportunity to co-design academic courses with academia and three great Australian universities to produce better outcomes in a more efficient manner while supporting planning for the future of professional military education—a better integration of outsiders into continuums, complementing the JPME continuum that was endorsed in 2019. We've received input from every senior military leader on specific areas of interest for students in their papers, and we've established different mechanisms for students from across our public service, our military, and industry to participate in courses, and also some of the electives such as our peri group which uses science fiction to think about the future. For several years, we've taken students on long and short courses from both federal and state agencies as well as different universities and the media, and we are including more students from different large industry groups this year. It's a start but will require more students from beyond the military if we're really to exploit the intellectual potential that a deeper collaboration between the military, academia, and the industry offers.

And finally, the development of a learning culture. I'd describe our learning culture as a work in progress. The Australian defense force is certainly not anti-intellectual. Given the range of internal training education institutions it possesses and its collaboration with academia and its partners overseas, such as the United States, this would be an unfair characterization. That said, it does not always learn as a totally joined-up organization nor are many of the incentives in place allowing for the full exploration of different ideas that will be possible in a democratic society. Our promotion pathways are generally traditional and lately have become centered on the capacity to work the strategic committee system and operate with other agencies in our national capital, Canberra. My sense is that perhaps the Russian senior officer promotion system went the same way a few years ago, but this is a serious issue because to develop a learning culture that gets the best from a collaboration between the military and its industry and academic partners requires military officers who also deeply understand war, competition, and warfighting. These are the things that set us apart from what no one else can do. We must have this special knowledge as a foundation in the three-way collaboration as I discussed in the presentation.

To conclude, I think there's much that might be done in my own country and potentially in the United States to develop and utilize a closer relationship between the military, academia, and industry. Besides the opportunities resident in our own individual nations, there are other international partnerships we might exploit to spur the relationship with the military and its academic and industry partners. Both the quad and orcas have immature conceptions of cooperation beyond very narrow bands of strategic and technological areas. These could potentially be broadened to develop our own 21st-century intellectual arsenal of democracy which incorporates not just the best equipment with the best ideas and the best training, education, and educated people. After all, there are no silver medals in the profession of arms, and second prize in the grand strategic competition of the 21st century is a very unattractive proposition for democracies, but perhaps Wick Murray put it best when he wrote that war is neither a science nor a craft but rather an incredibly complex endeavor which challenges people to the core of their souls. It is, to put it bluntly, not only the most physically demanding of all the professions but also the most intellectually and morally demanding. The cost of sovereignly thinking at every level of war can translate into the deaths of inerrable men and women, most of whom deserve better from their leaders, and that would be us.

Now I would propose to you that now is not the time for slovenly thinking or even normal thinking. It is a time for clever and connected thinking about the challenges ahead, and this includes a closer relationship between the military, academia, and industry. Thank you and I'll be happy to take any questions you might have.

8A

Leveraging Higher Education to Grow Military Strategists

Panelists:
- Nicholas Murray, Ph.D., FRHistS, Adjunct Lecturer Wargame Designer and Instructor for the Strategic Thinkers Program
- Colonel Francis Park, US Army, Ph.D., Director of the Basic Strategic Art Program at the US Army War College
- Dr. Robert Davis, Associate Dean of Academics at the Command and General Staff School

Moderator:
- Ken Gleiman, Ph.D., US Army (Retired)

As presented at the 2022 Strategic and Security Studies Symposium
Hosted by the Institute for Leadership and Strategic Studies
University of North Georgia

Dr. Ken Gleiman:

Well, good morning, everyone here for our first panel of the day, I'm Ken Gleiman the President of the Army Strategists Association. We were proud to have a role in planning this symposium and we're really happy to be sponsoring this panel because we're going to talk about our favorite topic, which is military strategy and military strategists. But particularly, today we're talking about the role of higher education in educating and developing military strategists. I'm pleased to be joined today by three outstanding people, each of these people with me played an important role in my own career and my own education in military strategy. Now, Colonel Francis Park is joining me here at the table, while Robert Davis and Nick Murray are joining us virtually, and I'll introduce them each separately as we get to their part. They're here this morning as representatives of three premier programs that help to educate and develop our nation's military strategists, strategic planners, and strategic thinkers.

I use those three terms deliberately as they are either part of the title or the objective of these programs that we're going to highlight, and I'll emphasize it may also be good to think about what those terms mean and how they are similar and how they're different. So, for the past 30-plus years, the US military in general but particularly the United States Army has been on a quest to educate officers as strategists, strategic planners, and strategic thinkers. If I had to pick a point of origin for the story of that quest, I think we need to go back in the way back machine to 1989, the year the Cold War was ending, when Francis and I were just seniors in high school contemplating careers in the military and thinking about being cavalry or infantry officers and thinking about all the glory that would bring us. But it was that year that General John Galvin was in the middle of his second assignment as a Combatant Commander; he commanded US SOUTHCOM from 1985 to about 1987, then he became the Supreme Allied Commander in Europe from 1987 to 1992.

In 1995, he wrote an article in the Army War College publication *Parameters* entitled "What's the matter with being a strategist." In the article, General Galvin uses some very nuanced academic language to explain why he thinks the Army and the US military might need to consider educating strategists. Now I'm going to quote from that article, so see if you can read through the nuanced esoteric language he uses as I quote from the very first paragraph of the article:

WE NEED STRATEGISTS!!!... in the Army and throughout the services at all levels. We need senior Generals and admirals who can provide solid military advice to our politica leadership. And we need young officers who can provide solid military advice, options, details, and the results of the analysis to the generals and admirals. We need military strategists' officers all up and down the line because it takes a junior strategist to implement what the senior strategist wants to be done and it usually takes the input of juniors to help a senior strategist arrive at his conclusions.

When he was Chairman of the Joint Chiefs of Staff, Admiral Bill Crowe said, what we need are people who can deal with a "thorny problem, people in uniform who are expert in their warfighting specialties and also able to assist the national command authorities in matters of strategy, policy, resource allocation, and operations." These officers, he said, "need to be tested leaders, skilled military technicians who are open-minded and adaptable, knowledgeable of military history and the role of armed forces in the world and versed in the complexities of bureaucratic decision-making and the international interests of the United States and their allies."

Now, of course, I guess there's absolutely nothing nuanced about General Galvin's point. He was putting the bottom-line up front and you can almost sense his frustration, and I sometimes wish I could teach my students to get their bottom line up front in their essays the way he did, although he may have overdone it there.

But imagine his predicament, imagine being the highest-ranking military officer in Europe, just as the world you've known throughout your 40-year career is about to change and get a whole lot more complex actually—to be fair, the world was always complex, but you're just starting to feel it and now you're trying desperately to understand it. Well, I'm not so sure that the Army or the other services heard General Galvin's plea because the article was reprinted in *Parameters* in 1995, right about the time I was freezing my butt off at the mountain phase of ranger school thinking about how with God as my witness I would never come back to Dahlonega, Georgia again. And I have to say it's so much nicer on this side of the fence and with food, but at that time the military was dealing with the civil war in the former Yugoslavia and getting ready to think about implementing the Dayton Accords and, again, there was this need for strategists.

Over the last 25 years, the Army and the services eventually got around to really answering General Galvin's call with many different programs, all of which leveraged higher education. In 1998, the Army created a strategist functional area and went by a different name but sort of morphed and then, by 2003, had launched this program called the Basic Strategic Art Program or BSAP. In 2012, the Army created the Advanced Strategic Planning and Policy Program, or ASP3, which is also known as the Goodpaster Scholars, named after General Andrew Goodpaster, who is actually a fascinating story of a sort of self-made strategist. And then in 2016, the Secretary of Defense, Jim Mattis, somebody who most of you are familiar with, created the Sec Def Strategic Thinkers Program (STP), out at the school of Advanced International Studies at Johns Hopkins.

Each of these programs leverages higher education to educate and grow future military strategists. These three programs that we're going to talk about are by no means a comprehensive list of all the efforts to raise strategists or improve strategic thinking across the military, and they're not the only way that the Army or the services have addressed, the demand signal in Galvin's article. But I think they do represent some of the most meaningful efforts where these initiatives, in fact, had the buy-in of senior leadership and or a very determined champion within the bureaucracy. And my hope is that, as we outline these three programs as case studies in the education of strategists, strategic planners, and strategic thinkers, that it'll spark some questions and ideas about how we can leverage higher education further for the future to educate strategists.

So first I'm going to introduce Colonel Francis Park. Colonel Park is the director of the Basic Strategic Art Program at the US Army War College. After commissioning in 1994 from ROTC at Johns Hopkins, his first 10 years on active duty were spent primarily in armored cavalry and light airborne cavalry assignments before he became an Army strategist, and he was actually in I believe the third cohort of the Basic Strategic Art Program of which he is now the director. Since 2004, he served in operational and institutional planning, strategy, and policy assignments ranging from all the way down to the division level and all the way up to the Joint Staff. He's a principal author of many strategy documents that many of you have read as well as of documents that many of you have not read or probably will never read because they were buried—which is always a great beer story with Francis about why some of his work is hidden. Colonel Park holds a master's degree from the School of Advanced Military Studies and a Ph.D. in history from some school in Kansas that's apparently good at basketball, but not much else. And, most importantly, he does hold the highest award that can be given to an Army strategist, which is the Order of Saint Gabriel Gold Award, and he's one of only two recipients of that award. So, without further ado, I'm going to turn it over to Francis.

Colonel Francis Park:
[Please see peer-reviewed article entitled "Educating Strategists: An Introduction to the Basic Strategic Art Program" in this collection.]

Thanks very much. It is my great pleasure to be with you this morning. I will talk a little bit about the program that I run and the method that goes around it. I do need to include the obligatory disclaimer that the views presented are those of the speaker and do not necessarily represent the views of the Department of Defense, the United States Army, or the US Army War College. So, with that out of the way, this is the purpose of the course:

> Provide officers newly designated into Functional Area 59 (Strategist) an introduction to strategy and to the unique skills, knowledge, and behaviors that provide the foundation for their progressive development as Army strategists... Create a shared common foundational experience, acculturate officers to the functionalfunctional area, and assist in the creation of their FA59 self identity [and] build the FA 59 network.

The thing that I would really call out from the purpose are the skills, knowledge, and behaviors, skills in as much as what are the things that we need our strategists to be able to do? What are the things that we need our strategist to know? And how do we want our strategist to think? The other thing that BSAP provides is it's a common cultural experience for all Army strategists regardless of components, whether in the active Army, the Army national guard, or the Army reserve. Also, it provides a single central experience for all strategists. Prior to that, there wasn't one, and that acculturation is actually quite valuable. And if there's any one recurrent theme that I've heard both this morning from General Ryan's comments and from the speakers yesterday, it is culture; culture is a recurrent theme, and it's one that we ignored to our peril.

So, some conceptual foundations. The foundations for development—and I'll throw this out: my views on developing military strategists are a matter of public record. They showed up in an article that was published in *Infinity Journal* in 2015, but they rest really on the three foundations.

The first is civilian education; it provides you an intellectual basis primarily for the unknowns. Professional military education is the second one that provides you not only a military theory but also a professional basis for the practical application of strategic art out in the force. And then the third is relevant experience. It is one thing to have civilian education, it is one thing to have military training, but it's another thing to have mastery, and you only get that out in the force. So relevant experience is actually quite valuable. You may get some basics in school; you will get proficiency and eventually mastery and strategic art and repetitive assignments in the sorts of things that you're going to do—much as we wouldn't ask an infantryman to command a light infantry battalion if his entire experience had only been in mechanized forces prior to that, at least we shouldn't be. We expect our strategists to have a broad sense of experience across the competencies they might be exercising.

The application of those standards shows up in Department of the Army pamphlet 600-3, which lays out the requirements for professional development but also for military education level four, which are the standards for Staff College for an Army strategist. The first is a master's degree in a strategy related field, preferably from a civilian institution and, you know, General Brown's comments yesterday about what he learned at the University of Virginia come to mind and how he uses those skills every day. Among the reasons why civilian education is actually quite a valuable part of that is the civil-mil interface that was discussed yesterday. Part of that is also the diversity of thought and, while this is not necessarily like military PME conferred degrees, it's important to distinguish between an academic degree and a professional one. And I would posit that the degrees that are conferred in a US Army course are generally professional degrees. I had a very different experience at the school of Advanced Military Studies than I did as a part-time student as a Lieutenant taking master's degree courses when I was at Fort Hood or compared to my experience as a full-time graduate student at the University of Kansas in the history department there. It is valuable to have something that is quite different from the military because many of the strategists we have are going to interact with civilians who don't necessarily speak in what I derisively call "defense speak."

The defense strategy course is a 19-week correspondence course that prepares strategists initially for the War College but is also being used as a prerequisite for BSAP. They also attend the intermediate level education common core typically at a satellite location, although there are some who attend at Fort Leavenworth, and then the Basic Strategic Art Program is the last of those four requirements. We apply those standards regardless of components; the phrase is one federal standard because our Army Guard and Army Reserve strategists are held to the exact same standard as our active-duty strategists, and today is unusual in that I'm wearing a uniform. Most of the time we're in business wear; I do that, or the program has done that, to break down barriers amongst the students but also because we don't really care what component you came from; we don't really care what rank you are; we judge you by our standards, and we are a very standards-based organization.

There are three assignment categories in a functional Area, 59: institutional, operational, and applied. Institutional very broadly is how you build a future force. Operational strategy revolves around how you employ the force; both of those activities can include not only strategy development but also planning and strategy implementation. Applied strategies are the third category, and those are other assignments that require foundational skills that are institutional and operational education; and commander's action groups, for example, fall in that category; interagency assignments also follow this because strategists need a basis in institutional and operational strategy to be able to represent the Army or the Department of Defense in those organizations.

There are a couple of observations that I've made over the last 18 years of being a strategist. One is that the skills that strategists learn across planning, resourcing, strategy, development and policy— those are complementary and the unity is greater than the sum of its parts. And as you, as strategists get more and more senior, they will employ all of them in concert, which requires development across all the assignment categories. We'll get 59s typically about the seven or eight-year mark and, given the usual timelines to get through all the schooling, we get them at about nine or 10 years into the Army, which gives us about three to four jobs before they're in the primary zone for Colonel, which means there's some career planning that has to happen fairly early on.

The other thing, and I will speak to this from personal experience, Colonels are a little slow to learn new skills, and old dogs don't learn new tricks very well. So, if you're going to teach any new tricks, they better learn them as majors or lieutenant colonels, because I have seen Colonels try to learn new tricks, and it's fun to watch but not fun to experience. So, a couple of key points about the course; BSAP is the foundational course to transition basic branch officers, whether armor officers such as myself or infantrymen. The majority or the highest population of functional area 59 officers are infantrymen, but we have representation across almost all the branches. BSAP began in 2003, expanded to two classes in 2006, and now, runs three classes a year. We have 16 students per class for a total of 48 students a year. We conduct instruction up to and including the secret collateral. We are a practitioner course taught by practitioners for practitioners, which means that we adjust the course on the fly to account for feedback from the field. The 59 proponent at the headquarters Department of the Army is on speed dial for me. I talk to him regularly, but the reality is that our graduates need to be ready to fight upon arrival to their unit, and that's usually a combatant command, or an Army service component command, or a joint task force. So, our sweet spot for functional area 59 officers is at the three or four-star level.

We do have academic attrition; it has gone as high as 25%, but it is generally for reasons of writing and critical thinking skills. Attrition has leveled out to about one per course either for administrative reasons or for academic reasons, and that's a pretty hard wake-up call for some officers who show up thinking that BSAP is like any other Army course—I guarantee you it is not. And between the faculty—and you can see the credentials; we have a combined 59 years (which is a coincidence; I did not come up with that) of experience in the functional area. And all of the faculty have served as division chiefs at some level or another, typically at the national level or, in my deputy's case, at the US Army Special Operations Command.

So, this is the course. It's six sub-courses; it is largely the same structure that existed from the very beginning, but I'll walk through the sub-courses and talk a little method to the madness. Strategic theory, you know, we start with that; it is always the first block. We provide the theoretical foundations for how strategists need to approach their discipline. We follow that by strategic art, which are historical case studies that teach how the theory is applied. But also if they illustrate things in the absence of hard experience, history is probably your best substitute to find out how to learn from somebody else's mistakes. Contemporary strategic challenges are a guest speaker program. We normally invite experts in the field on concerns of current interest to the joint force and that provides the context for things they're going to do later in the course. National security decision-making is going to be familiar terrain for a lot of political scientists; we do policy formulation and decision-making at the national level. The last two relate to institutional strategy and how you build the future force? And then joint Army planning, which is really focused on the combatant command of the theater Army. How do we employ the jointly empowered force that we just built? Bottom line is that you know we teach strategists how to lead multi-disciplinary groups, and senior leader decision-making strategy is inherent across all instruments of national power, and planning is inherent across all elements of combat power. So, all of our graduates need to have at least some fluency across all of those.

So, these are the expectations that I would tell people when I briefed this slide to leaders out in the force. This is what you can expect of one of our graduates; they're going to have the intellectual, theoretical, and professional foundations in policy advice, strategy development, and operational art. This prepares them to work from the core level to the national level, which is where our assignments happen to be. We expect our strategist to be able to lead operational planning teams in not only design-like processes but also deliberate planning processes to write orders. We expect our strategist to be able to develop strategies and plans, whether theater, strategic, or national strategic objectives. And if somebody had told me 18 years ago that I might find myself writing the national military strategy, I would have laughed really hard; that laughter stopped in 2017 when I found out that I had to rewrite the national military strategy. So that was the story of my 2018. I don't remember much of that year, but I do remember leaving the office late, which was not by design. We do expect strategists to be able to know what's the institutional Army, what's the joint force, what's the interagency, how do we apply that? And how we use the force, how do we refine how the force currently exists, and how do we develop a force for the future against potentially the unknowns? And then we expect our strategies to be effective communicators who can write, speak, or visually describe what it is they're trying to do in an effective manner for senior leaders. Our model is, at the bottom, fortune does favor the prepared, and our mission is to get strategists to that point so that they can be ready to fight when they get to their units. Thank you.

[See Appendix for corresponding PowerPoint presentations.]

Dr. Ken Gleiman:

We're going to go to Nick Murray who is representing the SecDef's Strategic Thinkers Program at Johns Hopkins. So, Nick teaches and runs war games for the SecDef's the second Strategic Thinkers program out of Johns Hopkins as well as actually part of the team that created that program. He has designed and run more than 100 war games. In addition, he is an active scholar with numerous books and articles, including several on the topic of professional military education. So, it's a fantastic thing to sit down with Nick and talk about PME because it is one of his favorite subjects. He's advised and assisted the Office of the Secretary of Defense with the policy regarding military education and war gaming. And he's received numerous awards, including OSD's highest medal, The Exceptional Public Service Award. So, Nick, over to you. Good to have you here virtually.

Nicholas Murray:

Good morning. I'd like to say thank you to the University of North Georgia for facilitating this conference. It's an important topic, and I think it's one that is not as thoroughly examined in the services as it perhaps should or could be. Although there are periodic looks at it, where people throw their hands in the air and say what's going on? Why can't we do x, y, or z without necessarily doing a more thorough examination of some of the issues at hand? I'd like to thank Ken as well for putting together the panel and organizing this.

I just want to stress that I'm here on behalf of the Strategic Thinkers Program (STP) at Johns Hopkins; I work with Professor Dan Marston, the director, who sends his apologies as, unfortunately he was already committed to another conference. I'd like to give some background, and I'm going to take a slightly different tack than I had originally planned, largely because of Colonel Park's and General Ryan's thoughtful comments. As such, I'm going to add in some things I hadn't originally prepared which relate to the context and background of the STP, and which now seem important to include. I hope you'll bear with me.

The STP was created in response to long-term professional military education issues identified regarding the development of strategists within the PME system. They have been consistently discussed over the last 50 years, and I think the three programs represented here are all attempts at addressing them. The House

Armed Services Committee (HASC) Report on Professional Military Education of 1989, better known as the Skelton Report (named for Representative Ike Skelton of MO), looked at the way the services educated their officers, what the services were doing well, and perhaps were not doing so well. Many of the findings were repeated in the HASC Report on PME from 2010, but I will focus on the former report as that is still fundamental to this day. One of the key things the Skelton report said the services were not doing effectively was the production of strategists: or strategic planners, and strategic theorists as they named the two sub-fields. The report "identified three major components in the development of a strategist—talent, experience, and education." The report further concluded that "the selection, assignment, and education systems need to be better coordinated... to maximize ... these three factors." (HASC Report 1989, p. 28)

The report found that strategists (of whatever ilk) needed to be analytical, pragmatic, innovative, and broadly educated. The report further recognized that only "some" would possess all of these attributes at a sufficiently high level, but sufficient officers could be developed through exposure to relevant experience and education. (HASC Report 1989, pp. 27-31) Thus, only limited numbers of strategists of all types were needed, along with a smaller group of "strategic theorists" being likely or necessary.

[T]he goals of the PME system with respect to strategists should be two-fold: (1) to improve the quality of strategic thinking among senior military officers and (2) to encourage the development of a more limited number of bona fide theoretical strategists. (HASC Report 1989, p.28)

Talent alone was not considered sufficient, and officers required appropriate experience and education as well as a high degree of proficiency at the tactical and operational levels as well as within the Joint force. "Officers who are intelligent, imaginative, articulate, and interested in studying strategy must be identified as early as possible during their careers so that their development can be facilitated by appropriate personnel policies." (HASC Report 1989, p. 29) The select officers needed to be generalists, and the core subjects to be studied were: history, international relations, political science, and economics. (HASC Report 1989, p. 30)

Thus, over thirty years ago Congress set out its vision for the development of a core small group of strategists. Although the Services had done well with the development of strategic planners, it had not done sufficiently well overall to meet the intent set out in 1989.

The topic of the development of strategists came up repeatedly, and far too numerously to mention here so I will stick with the main Congressional Report on PME from 1989. The subject came up again in 2010 (HASC Report 2010), as well as in internal reports within the DoD and services. There was an identification that many of the existing programs were to a certain extent, and I'll give you an example of the language of the internal discussions, 'sinecures for a good old boy network,' or that the programs had essentially atrophied, and they were largely not doing what they'd originally been designed to do. So, there was an effort both to reform those programs, which is bureaucratically very difficult, but also to encourage the services to push and develop new programs and new ways to improve, not burdened by past performance or bureaucratic failings. That's a big picture background, if you will, in terms of the service side and the push from senior leaders to set up and develop the kinds of leaders and thinking needed to wrestle with the types of complex problems which seemed to be emerging.

Dr. Gleiman mentioned that Secretary of Defense Mattis signed off on STP's creation in 2018, and while true I think this is part of the issue that General Ryan talked about. That is, even something relatively small like STP required the SecDef themself to advocate for and approve before it happened. It is, thus, important to note that STP did not spring forth from nothing. STP, in one form or another, has been in existence, to a certain extent, since 2010 when Professor Marston created the original Art of War Program at Army CGSC. His program worked in conjunction with the Master's Degree Program in Military History, which I happened to direct, and the requirements for the two merged soon after: in terms of the Art of War students. He moved to Australia in early 2013 to create a similar program there, and that summer I was asked to brief the idea to the OSD(P&R). From there it started to work its way through the Pentagon. The concept was written and re-written, and briefed, over and over again; it took actually more than four years, and several SecDefs pushing the concept, to get to SecDef Mattis's desk. So, we were fortunate; we had three SecDefs in a row who were all very supportive, but it still took time.

Despite the call from Congress and the SecDefs, as well as multiple three- and four-star senior leaders across the joint force being supportive, it was stalled. Typically within the mid-level of the bureaucracy within the DOD in particular and to a certain extent within the Army, as well and other services. The Army was problematic, in particular—the bureaucracy there: please note CGSC had helped set up an early version, and they did a very good job with that. It was stymied for years until multiple senior leaders came along and said this *must* happen. It was only then that it was pushed through the system at the end of 2017. That was when the final memorandum establishing the formal call for the program was signed. So, SecDef. Mattis should receive credit, but I think it took, essentially, three separate Secretaries of Defense, as well as several Chiefs of the General Staff, and multiple senior leaders to intervene in the process. This is where the story of STP ties in with General Ryan's comments about things being done in a timely fashion, the role of organizational culture, and an openness and willingness to accept and try out new things. Even with the support of senior leaders, this is a process that took actually multiple years to put forward, and I think that's worth repeating.

With that in mind, one further push came via a specific direction from the National Defense Strategy (NDS) to improve intellectual leadership and military professionalism within PME. This is where luck also played a large role. If you are familiar with my work, you will know I am a fan of Clausewitz's writing and ideas, and especially that the role of luck is central to his understanding of the nature of war. As such, I wish to add one last point. On the day when the memo regarding STP went to SecDef Mattis for discussion, it was accompanied by an article on academic rigor in PME[1], an internal memo on wargaming I happened to have written, and another memo on joint billeting I had assisted with. I also happened, by sheer luck, to be in the building on the same day and at the same time, and was whisked away from my appointment to meet one of Mattis' aides to discuss STP, wargaming, and joint billeting. So, despite all the planning, the years of preparation, and the high quality of Professor Marston's original creation, it all required a large slice of luck to come together. And that, I think, supports General Ryan's argument.

With the above now out of the way, now to STP more directly. STP exists as a means to develop strategists and as a complement to other programs—that's the thing, I think that sometimes programs have been seen to be in competition with each other rather than complementing each other and that resistance to them sometimes stems from this. I think that's an important point because programs like STP and the one Colonel Park is representing, the ASP3 Program at Army, complement each other, and they're not necessarily in competition for the same officers. The services select the officers they think would be best suited to each program.

Now I will explain the reason for going to a hybrid sort of PME at a civilian school: that came directly out of the House Armed Services Committee Report on Education in 2010. One of the key ideas was that, if we're going to have strategists they

---

1 https://warontherocks.com/2016/02/rigor-in-joint-professional-military-education/

need to be exposed to civilians and civilian decision makers—this idea includes many of the things that Col. Park and Gen. Ryan have already mentioned, so I won't go back over them—but it is important to give access to select officers to the civilian education system. This is both to expose them to civilian ideas and norms, as well as vice versa. Those kinds of experiences are not always available within the PME system, for a number of reasons. Across the PME system, there is a great deal of excellence, and I think that that's worth emphasizing. The downside is there are often pockets of excellence that frequently don't talk to each other, cylinders of excellence, as my officers like to refer to them. And that these cylinders themselves are often small when it comes to the subjects listed as being key to the development of strategists. So, I think one of the things that civilian schools offer is, one can have a large department or multiple departments with expertise in a broad variety of different fields all in one institution. Within PME schools, that is actually very difficult to achieve. Often, as I have mentioned, there is genuine excellence in one or two areas, but that excellence is typically small and limited in any one institution. That is one of the things that, the House Armed Service Committee and the DOD were looking for, and we were put forth in our proposal for STP. Dan and I felt that certain civilian schools were better placed to teach the broad range of subjects and historical background called for, and could offer the students both the military educational side to tap into their experience and the broader civilian version of that, with multiple different civilian experts on site teaching a broad range of courses.

In line with the HASC 1989 and senior members of DoD, Dan and I thought the above combination would provide an incredibly good foundation. In addition, as both General Ryan and Col. Park have mentioned, the diversity of civilian schools is an important part of the experience. Furthermore, civilians without military education get the benefit not just of providing different ideas but also of receiving ideas from service members through being educated with them. I will provide you a quick example, and I think this ties in with the benefits of programs like STP and the Army's other service's programs as well. I ran a war game at Saint Andrew's about 18 months ago, but it was pre-COVID, so I'll just put it like that because there's that whole era lost to that thing. What I found was the civilians came up with some absolutely brilliant ideas in terms of running the war in the Pacific from August 42 onwards at the strategic level. They also came up with some things that were completely infeasible and fundamentally stupid. Now one can't just say to a student that's stupid, you're a moron, don't do it again. Well, I suppose one can, but that is another conversation. What we found was the civilians had an incredibly broad range of wonderful ideas, some of which were 'please, don't go there' all the way through to, 'well, that's absolutely fantastic'. My military officers frequently might not have gone to some of the sublime ideas put forward by the civilians because the organizational culture that the service members have grown up in would not allow that. This ties in again with that discussion of risk-taking that Gen. Ryan mentioned in his talk early on. Cultural norms wouldn't always allow that kind of risk-taking sometimes even within the context of a war game. So, a civilian with no frame of reference can go there because they don't have that concept of the difficulties of implementing their idea. But the civilians also saw the benefits of the connection with service members. There was also an appreciation of the need for military experience that Colonel Park mentioned to frame an idea. Some things simply aren't feasible and can't work for technical reasons, and that might escape someone without a military background. So, the cross-over benefits of the interaction between civilians and officers is an important part of having STP at a civilian school. If strategists are required, and I think they are, officers have to be exposed to the very people who are likely to be developing the policies the officers will be tasked with implementing.

Now we wanted to have the program at a 'Top Tier' school. This was also one of the requirements of the House Armed Services Committee Report from 2010. Furthermore, the school needed to be one that understood the needs of the service members and DoD. Now that provided a fairly limited number of schools which fit the criteria. The term 'Top Tier' was never actually defined. So, we used that definition in a broad way to speak with individuals within civilian schools to get feedback as to how the whole thing might fit together. That meant reaching out to personal contacts, as well as people who had experience of dealing with DoD. And that's the other part of the creation of the program: personal connections actually play an important role.

How we have set up STP, and please remember this is Dr. Marston's original creation. He, created, essentially, a core course which is made up of four parts, along with two wargames and a capstone staff ride that form half of the master's degree in international public policy at Johns Hopkins SAIS. Students are free to take any of the other classes within that program. Now some of the students will have service requirements that require something like "a couple of classes with an emphasis on China," but what we've found across the range of students is they are able to take a really good broad range of master's level classes from top experts. They can then bring back their learning into their service experience with their following assignments.

How does that fit in with the DOD more broadly, and what does the course look like? I'll give you an idea of those two things. How it fits in is that it takes place roughly at the ILE, the Intermediate Level of PME for officers or after, but typically before the Senior Level of PME.

Colonel Park mentioned that when an officer gets to the rank of o-6, sometimes it's too late for them to learn new things, and I'm in full agreement. I used to joke with my officers that once you become an o-6 you've been ruined, because if you're not making General, you're going to be bitter, and if you are making General, you've got to practice that horrific attitude that often comes with senior rank. All kidding aside, by that time officers have been in an organizational culture for typically 16 to about 22 years, which means that sometimes ideas from outside can be very difficult to take on board. Something that is the case for any organizational culture.

So, having the officers come in typically at the rank of 0-4 to 0-5, there are some 0-6s, seems to be beneficial in terms of preparing them for the next couple of stages of their careers. STP has focused on rigorous intellectual debate, intellectual discussion, and communication. We heard from General Ryan and from Colonel Park, that the ability to communicate is one of service officers' biggest failings; it's often one of the areas where students will trip up. I found that in running the master's degree program for what is now Army University, the ability to communicate is both vital and also often underrated, and it doesn't get sufficient traction within the services because it's very difficult to do. It is incredibly time-consuming, and requires a great deal of iteration to improve. This is why writing and effective communication form such an important part of the mechanisms of evaluation in STP.

The way the STP works is students meet twice a week; they do four 14-session classes in the year for a total of 56 sessions. The sessions are four-hour seminars, and students are charged with answering a question and making an argument with a 1,000 word essay every other session. That argument is disseminated to the other students, and the students who are not answering questions for that session write a 300-word rebuttal. That is the plan for every class for all 14 sessions, and that is repeated for all four of the core modules within STP.

The students function as their own white team, if you will, and they'll actually police and hold each other to account. It is, effectively, a slightly modified version of the Oxford tutorial method. The key is to ask questions to draw out ideas and then encourage the students to have a rigorous debate—and sometimes that can be not quite coming to blows—but really quite aggressively argumentative. Between the extensive writing requirements, and vigorous debate, is something that is incredibly beneficial in terms of the ability to communicate: which is a vital part of the job of a senior leader. They are evaluated on their ability clearly to communicate and make arguments in the context of connecting the choices and the actions of the leaders and conflicts they are studying. This structure is used for classroom discussion, staff rides and wargames. Students need to show the connections between the choices and policy, within the political context of whatever it is they're doing. So that's how they are evaluated.

The practical part of the course: there are two wargames: one just before the Christmas holiday break and then in the middle of March, where they take part in a strategic level wargame. For the one in December, they just did a war game I designed looking at August 1942 into July 1943 in the Pacific theater. The students are given the role of one of the theater commanders: typically the Japanese Army commander, the Japanese Navy, Nimitz, MacArthur, Blamey, etc. They are given the original historical guidelines and policy, and then they have to actually make the best of their actions from that. And what they often find, as in with real life, is the policy guidance doesn't match the means that they have at their disposal, neither does it necessarily match the tasking they've been given, or the resources, and yet they still have to unscramble that and craft something that makes their joint force effective against their opponent.. And that includes individual members within a team having different goals because of their guidance actually pulling them in a different direction. A large part of the wargame is the negotiations within the team, as well as actually connecting their actions to the broader policy goals.

The capstone is a staff ride where the students are given a campaign theater where they then have to identify stands, key locations, connect governmental policy through the levels of war down to the tactical level and back up. They have to explain and discuss the decision making and choices of the protagonists and demonstrate the connection to a greater whole. This year the staff ride followed British Fourth Army in France in 1918, which included the roles of British, Canadian, American, and Australian units working in a Joint Forces. I hope this has been helpful. Thank you.

Dr. Ken Gleiman:
Okay let's go ahead, and if you can put those slides up, I will give a quick five minutes on the Goodpaster Scholars ASP3 Program. So I wasn't really prepared to do this, but I am familiar with the program because way back in 2012 as I was coming out of Afghanistan, I got a call and was asked if I'd like to be part of it or if I wanted to apply for a new Army Ph.D. program, and I had not much of an idea on it other than that it was run by the School of Advanced Military Studies. But the idea essentially for this program came from General Ray Odierno, who passed away this year, who was a senior commander in Iraq during some of the most difficult days in Iraq. And one of the things that he realized he needed to do was that he needed help understanding the complexities of that situation, and he found himself and his strategists on hand and military planners reaching out to academia, reaching directly out to scholars at universities as well as think tanks. And that helped him frame the situation in Iraq that eventually helped to the surge, and that's a story that I think is well written, but when he became the Chief of Staff of the Army, he said, "I want something more. I want a new program where we are going to build solid connections within the Army to academia and to institutions of higher learning." And the program they came up with was the Advanced Strategic Planning and Policy Program, and the idea behind the program was to somehow send about 12 officers a year to fully funded Ph.D.

programs. Why did they go with a Ph.D. program? Well, they believed that, as many of you have Ph.D.s know, part of getting a Ph.D. is doing the research, searching for answers to complex problems and essentially creating new knowledge. And I believe that a lot of the problems that we face at the strategic level require that level of time and energy of creating new knowledge. So, they decided on a Ph.D. program. The problem with the Ph.D. program has always been time, and as I think Francis mentioned, there's a great deal of difficulty in time in an officer's career, because essentially if you're going to put an officer in a Ph.D. program, you are taking them out of the operational force for a significant amount of time. At a minimum, you're taking them out for three years, which is usually two years of course work, and then if you're lucky, you're good, and if you're super-efficient, you might be able to complete your dissertation in a year. So they developed this program so that they could select officers from across the force and get them to universities to study topics that dealt with national security and originally as they had conceived it, there would be two years of course work at a university, followed by an operational assignment, followed by a year sabbatical. It has generally followed that model. But a lot of times now we are actually taking officers out of the force for about three years, and they're given that amount of time to finish their dissertation. And then these officers are assigned across the force; they usually find themselves actually working with Army strategists, the fa-59s that Francis was talking about. In fact, Army strategists actually make up about 20% of the officers that do this Goodpaster's Fellowship.

I'm trying to think of what else I can say about the program, and I'm really sorry to Dr. Davis for not doing it more justice. I would say that, since 2012 with 12 a year, there are over 100 officers enrolled in the program, and over 50 of them have completed their Ph.D.s and are now serving all over the force. There is actually an article in *Military Review* that was published in 2012 that explains the origins of the program. It's a very worthwhile read and, as you can see, that's actually General Odierno there, and it kind of emphasizes this strategic planning problem, so to speak. That's the first cohort; a few of those folks are retired. The gentleman standing in the middle is actually the CAG Director of General Millie, so he's General Millie's sort of chief on-hand strategist right now, Joe Funderburke, a good friend of both of ours. And there's a fellow in the back there who's actually leading the joint staff J-5 plans right now. So, the program's done well in placing these officers in really high strategic positions. I know Robert had in mind these quotes, I think they're probably very effective.

The first quote: Bernard Brodie wrote in *War & Politics* that

> As one who helped set up the US National War College by serving on its faculty for the first year of its existence, and who later served on its Board of Advisers—as well as having given lectures there and at the other war college—I feel I can say with confidence that the training afforded at that level is by no means adequate to the needs described. It is undoubtedly a valuable training, and visibly raises the horizons of the officers who pass through it; but as far as changing their basic attitudes is concerned, the training is too brief, too casual, comes too late in life, and keeps the military consorting with each other. [486]

The second quote: COL(R) Harry P. Ball, PhD, wrote in *Of Responsible Command: A History of the US Army War College*, that

> the War College cannot take all credit, nor does it need to accept all blame, for graduate performance. A year of War College work is too short to produce judgment, character, and virtue where none existed before, and it comes too late in an officer's life to change dramatically his personality and system of values. At best the bad can be dampened and the good reinforced. [496-497]

The second one emphasizes that this is a current cohort doing research. I will also say that the interesting thing about the program is the selection process is very difficult; of the people that apply, I'd say about 10 to 20% are selected. They do first six weeks or so at Fort Leavenworth where they learn to be a graduate student again and get sort of a military side of the education and strategy. And then they'll do at that again later. They also get classes on, like, you know how to apply to university and some strategies for completing your Ph.D. because you know that all Ph.D. programs have enough attrition themselves.

So one of the great things about this program is it leaves room for command. So, if you're an Army strategist FA-59, we have to step away from command, you won't command troops because of the assignments it requires, etc., but with the ASP3 program, you're getting that strategic education, but it's also leaving room in your career to command troops at the highest level, so we have several battalions and brigade commanders here.

[See Appendix for corresponding PowerPoint presentations.]

8B

EDUCATING STRATEGISTS: AN INTRODUCTION TO THE BASIC STRATEGIC ART PROGRAM

Francis J. H. Park
Department of Military Strategy, Planning, and Operations,
School of Strategic Landpower, United States Army War College

[underwent separate external peer review]

ABSTRACT

The qualification course for the U.S. Army's strategists is the Basic Strategic Art Program (BSAP), a demanding resident course conducted at the U.S. Army War College at Carlisle Barracks, Pennsylvania. The course provides new Functional Area (FA) 59 strategist officers in the ranks of captain to lieutenant colonel from the Active Army, the Army National Guard, and the Army Reserve the tools and perspective to "fight upon arrival" to their gaining organizations. Graduates serve in joint and Army assignments from levels ranging from corps headquarters to the national level. Bolstering the BSAP curriculum is the faculty's basis in practice, which comprises over 50 years of FA 59 experience to the division chief level. The eventual goal is the creation of a cadre of skilled practitioners in the application of planning, strategy, and policy advice for the Army, the Joint Force, and the Nation.

Keywords: applied strategy, institutional strategy, landpower, national security decision-making, operational strategy, professional military education, strategic art, strategic plans and policy, strategic theory, strategist,

The Basic Strategic Art Program (BSAP) is a graduate-level resident program conducted at the US Army War College at Carlisle Barracks, Pennsylvania. It is the only course of its kind in the Department of Defense and serves as the basic qualification course for officers entering the US Army's strategist functional area (FA), which is designated FA 59. The program provides new FA 59 officers the tools and perspectives to bridge the gap between their previous backgrounds in tactics and operations and the challenges of operating at the theater strategic and national strategic levels of war and policy. It also introduces the officers to the unique skills, knowledge, and behaviors needed as a foundation for their progressive development as Army FA 59 officers.

Through the use of theory, history, doctrine, exercises, staff rides, and official visits, BSAP examines the formulation and implementation of policy, strategy, and plans, reinforced with case studies. This method provides insights and illuminate the thematic continuities in strategy, the application of force, and the conduct of war. The principal focus in BSAP is on military strategy and landpower—with an emphasis on how the Army is structured and employed within the theater to achieve strategic objectives (Department of Military Strategy, Planning, and Operations, 2022). As a function of that methodology, the course seeks to break students of inappropriate tactical bias.

While other programs of instruction deliver instruction like BSAP, what makes it unique is its audience, the career paths they follow, and their unique knowledge, skills, and behaviors relative to the rest of the force. Colonel David McHenry, a former proponent officer for the functional area, described FA 59 officers as a "cadre" for the Army—

keepers of a core body of knowledge in campaign planning, strategy, and policy advice for the Army, the Joint Force, and the Nation.

Conceptual Foundations for Educating Strategists

The US Army has designated certain officers as strategists since the 1970s, based on the successful completion of a strategic studies elective track administered at the US Army Command and General Staff College since 1975. Completion of that series of electives—combined with a thesis-length research project and an advanced degree in a strategy-related social science discipline (often political science, economics, or public administration)—was required for conferral of the skill identifier. After returning to the force, officers with the strategic studies skill identifier were expected to alternate between command and strategist assignments, but in practice, command usually took precedence since promotion to colonel and potential general officer rank rarely occurred without command of a battalion or squadron as a lieutenant colonel (Park, 2015, pp. 09-10; US Army Command and General Staff College, 1977, p. 63). However, no formal career track for US Army strategists existed until the 1997 advent of the Officer Personnel Management System XXI, which enabled the creation of FA 59 a year later. Originally titled "Strategic Plans and Policy" at its outset, FA 59 was retitled as "Strategist" in 2010.

The first authoritative description of a US Army strategist owes to General John R. Galvin, whose career culminated from 1987 to 1992 as Supreme Allied Commander, Europe and Commander-in-Chief, US European Command. Galvin's 1989 description of a military strategist remains a concise articulation of what such an officer should be:

> A military strategist is an individual uniquely qualified by aptitude, experience, and education in the formulation and

articulation of military strategy (making strategy and articulating strategy are equally important). He understands our national strategy and the international environment, and he appreciates the constraints on the use of force and the limits on national resources committed to defense. He also knows the processes by which the United States and its allies and potential adversaries formulate their strategies. (Galvin, 1989)

The origins of "strategic art" owe to Major General Richard A. Chilcoat's 1995 monograph *Strategic Art: The New Discipline for 21st Century Leaders*. Chilcoat articulated a definition of strategic art as "The skillful formulation, coordination, and application of ends (objectives), ways (courses of action), and means (supporting resources) to promote and defend the national interests." In support of that definition, Chilcoat articulated three roles implicit to strategic art, namely those of the strategic leader, strategic practitioner, and strategic theorist. Each of these roles combine direct and indirect leadership with functional and conceptual expertise in developing and implementing strategy (Chilcoat, 1995, pp. 3–5). While Chilcoat's conceptualization did not see as wide acceptance as operational art, a concept that had become prevalent in the 1980s, it did gain enough traction to appear in the 1999 edition of the Army's leadership doctrine (Headquarters, Department of the Army, 1999) and lent itself to the 2001 creation of the Advanced Strategic Art Program that remains one of the most competitively selected electives in the Army War College curriculum. Its creator, Colonel Michael R. Matheny, would create BSAP two years later (Moore, 2009).

The career path and developmental requirements for a FA 59 officer appears in Department of the Army Pamphlet (DA Pam) 600-3, *Commissioned Officer Professional Development and Career Management* (Office of the Deputy Chief of Staff, G-3/5/7, 2020). It articulates a basis for development based on a 2015 article published in *Infinity Journal* (now the *Journal of Military Strategy*) that outlined a framework of civilian education, professional military education, and relevant experience (Redacted).

INSTITUTIONAL MANAGEMENT OF ARMY STRATEGISTS

The accession of officers into FA 59 entails a screening process for prospective candidates, followed by an initial entry education timeline of two to three years. The selection of Army strategists occurs through two paths. Prior to the current system, which started in 2012, selection of FA 59 officers had occurred through a centralized career field designation board held at Headquarters, Department of the Army. Prospective FA 59 officers were typically captains assigned to a basic branch who applied to enter the functional area after their company command or equivalent billet. If accepted, they left their basic branch and entered the functional area upon selection for major (Officer Personnel Management System XXI Task Force, 1997; US Army Human Resources Command, 2010).

The majority of new FA 59 accessions comes through the Voluntary Transfer Incentive Program (VTIP), which is the method by which officers change their branches or enter a functional area after commissioning into a basic branch. Most officers who elect to VTIP enter around the sixth or seventh year of active federal commissioned service. A small number of officers enter FA 59 via the Harvard Strategist Program. If accepted, an officer attends a one-year, mid-career Master of Public Administration degree offered by the John F. Kennedy School of Government at Harvard University, followed by a two-year utilization tour at the Army Staff, generally within the Office of the Deputy Chief of Staff, G-3/5/7 (Office of the Deputy Chief of Staff, G-3/5/7, 2020).

The FA 59 proponent has had a longstanding mandate that the initial qualification requirements for Army strategists represents one federal standard across the Regular Army, Army National Guard, and the Army Reserve (MacMullen, 2009), a standard that remains in effect to this day. As of 2020, officers seeking to become fully qualified FA 59 officers must complete the Defense Strategy Course (DSC), Command and General Staff Officer Course (CGSOC) common core, BSAP, and a strategy-related master's degree. The DSC is a four-month distance education program offered by the US Army War College that familiarizes students with concepts they are likely to see in BSAP or a senior service college. All officers must complete the CGSOC common core in either resident (fourteen weeks) or nonresident (generally sixteen months) form (Department of Distance Education, 2020). Another FA 59 qualification requirement is completion of a strategy-related master's degree, the discipline approved by the FA 59 proponent at the Army Staff. Finally, all prospective FA 59 officers must graduate from the sixteen-week BSAP. Consistent with the one federal standard, all of these requirements must be completed prior to consideration for promotion to lieutenant colonel, regardless of component (Office of the Deputy Chief of Staff, G-3/5/7, 2020).

Army strategists have served in assignments from the division headquarters up to the National Security Council. Currently, most FA 59s serve in 3-star and 4-star headquarters, whether Army or joint. The "sweet spot" for those officers is generally at the theater level, whether in a combatant command headquarters or an Army service component command. Strategists also serve in joint task forces, given the role those organizations have in connecting strategy to operations.

The current division of assignments in FA 59 into its three categories of institutional strategy, operational strategy, and applied strategy began in 2017. Each category involves the practice and application of strategic art pursuant to Chilcoat's 1995 definition (Office of the Deputy Chief of Staff, G-3/5/7, 2017). Institutional strategy entails the development of policy, strategy, and plans to build the future force. Institutional strategy activities correspond to the force development and force design horizons of the Joint Strategic Planning System (JSPS) (*CJCSI 3100.01E*, 2021). Operational strategy entails developing policy, strategy, and plans that account for foreign policy and the strategic environment to promote and defend national interests. Operational strategy activities correspond to the force employment provisions of the JSPS and integrate strategy with operational planning, execution, and assessment. Applied strategy comprises assignments that require the application of multiple strategic art foundations and depend on the combined exercise of institutional and operational strategy tradecraft. Many involve teaching assignments in professional military education institutions, or interagency or personal staff positions to senior leaders such as commander's initiatives groups (Office of the Deputy Chief of Staff, G-3/5/7, 2017).

These three categories of assignments provide a framework for proficiency and eventual mastery of the full range of strategist competencies. By design, there is no "golden path" in FA 59, meaning that officers have considerable flexibility of assignments, moving up or down echelons throughout a career. For example, as of the time of writing, the author's four prior assignments were to a 4-star joint task force, the Army Staff, a corps headquarters serving as a joint task force, and then the Joint Chiefs of Staff.

Strategists are expected to complete institutional and operational assignments prior to serving in an applied strategy role or prior to entering the zone of consideration for promotion to colonel. The rationale for the three categories of assignments and the explicit structural incentives to be a generalist are based on the mutually reinforcing relationship of those skills among the three. This relationship is especially true of applied strategy and its dependence on the foundations gained in institutional and operational strategy work.

Based on the timelines for officer promotion and the statutory requirements in the 1980 Defense Officer Personnel Management Act, a new FA 59 major has approximately ten years before entering the primary zone of consideration for promotion to colonel. In the absence of a "golden path," strategists must make an informed choice as to how they gain experience across the three categories. Within those ten years, a strategist can expect to serve in three to four duty positions. During Operations ENDURING FREEDOM, IRAQI FREEDOM, and FREEDOM'S SENTINEL, there had been more opportunities for one-year tours, generally as individual augmentees to joint task forces. The conclusion of combat operations in Afghanistan and the reduction of presence in Operation INHERENT RESOLVE have considerably reduced those opportunities, and they are now the exception rather than the norm.

The considerable demands for mastery for FA 59 colonels in supervisory positions makes learning new skills as colonels a painful experience for those organizations-turned-training aids, as well as the junior strategists in them who may be led by a well-meaning but ignorant senior strategist. Finally, the competencies unique to the three categories are not discrete for a senior strategist. Rather, FA 59 colonels should be able to synthesize competencies from all three categories, regardless of whether those positions are in service, joint, multinational, or interagency organizations.

PILLARS OF BSAP

One of the key learning outcomes of BSAP is that its graduates gain a rich professional perspective on policy, strategy, and doctrine through a solid intellectual foundation of theory, history, exercises, and staff rides. (Department of Military Strategy, Planning, and Operations, 2022) The primary rationale for that learning outcome is the basic competency for a strategist to lead multi-disciplinary groups and assist senior leader decision-making by assessing, developing, and articulating policy, strategy, and plans at the national and theater levels.

Based on that methodology, the BSAP curriculum rests on six courses, referred to as "modules," all touching on the core competencies for an FA 59 officer. Although some of the course modules have changed their names, the overall curriculum has remained constant since the first iteration of BSAP in 2003. The courses, while nominally discrete, are mutually supporting, and encompass the full scope of the likely duties an FA 59 officer will face across the three assignment categories.

The first of the course modules is *Strategic Theory*, which establishes a foundation in strategic and operational theory and provides students the theoretical tools to evaluate doctrine and strategy. Among the theorists examined in the module are Carl von Clausewitz, Sun Tzu, Mao Zedong, Thomas Schelling, and Hans Delbrück. In addition to its Army-common focus on landpower, the module also addresses theorists of seapower, airpower, and irregular warfare to explore potential aspects of continuity in national, cultural, or service styles of warfare.

The second course module is *Strategic Art*, which is based on historical case studies of military strategy and policy. The case studies range from the Peloponnesian Wars to the current era and focus on critical strategic themes to help explain victory or defeat. Among these themes are the correspondence of strategy and policy, theories of victory, mirror imaging, adequacy of strategy, prewar plans and wartime realities, coalition warfare, and civil-military relationships.

The third course module is *National Security Decision-Making*; the module acquaints students with the nature of US decision-making on national security matters through a primarily political science lens. It includes an examination of strategy development models, international relations theories, models for explaining state behavior, factors that influence strategy formulation and its execution, the national security organization, and policy development. Other topics include the theory and the practice of the US interagency process and analysis of the main actors engaged in national security deliberations. Students use Operation Desert Storm, Operation Iraqi Freedom, and other case studies that illustrate both formal and informal processes. Finally, students develop their own models for assessing strategy.

The fourth course module, *Contemporary Strategic Challenges*, is interleaved with other course modules at the middle of the course. This module combines assigned readings and guest lecturers to familiarize students with current and emerging contemporary challenges. Students will assess how these issues challenge US regional interests, objectives, and the resulting US policies and strategies. This module concludes with a comprehensive oral board presided over by BSAP and other Army War College faculty.

Looking inwards to the Department of Defense, the fifth course module is *Institutional Strategy and Planning*. This module familiarizes students with the organization, systems, documents, and processes of the Department of Defense, as well as that of the Department of the Army and its force structure. Using the force design and force development portions of the Joint Strategic Planning System as a framework for discussion, the module examines the structure of the Defense Department, as well as the processes and methods for envisioning future warfare, resourcing the future force within the years of the defense budget, and translating that future force into resources that are made available for employment by combatant commanders.

That examination of the future force leads into the final module of *Joint and Army Planning*. Oriented on the force employment portions of the Joint Strategic Planning System, this module examines the roles of the Department of Defense, the Armed Services, the Joint Chiefs of Staff, and the combatant commands to fulfill the ends of national strategy. Students study the rhetoric and practice of joint strategic planning, informed by national and combatant command level documents. Finally, the module familiarizes students with the processes and considerations for developing regional strategies, campaign plans, contingency plans, and landpower estimates.

Importantly, the curriculum is not static. It undergoes continuous adaptation to account for feedback from the field and factors in continued coordination with the FA 59 proponent to ensure that what is published in DA Pam 600-3 is fully reflected in the course. As a rule, a graduate of BSAP must be "ready to fight" upon arrival to their organizations, regardless of what echelon at which that graduate may serve.

Ultimately, this curriculum and its program learning objectives constitute a unity that is greater than the sum of its parts. Each of the course modules addresses an element of strategic art tradecraft that is mutually supporting to the other modules. For example, the author, when drafting the 2018 *National Military Strategy*, examined theoretical aspects of military power in space and cyberspace, for which there was little basis in doctrine or strategic direction at the time to inform the rewrite of the strategy. (Chairman of the Joint Chiefs of Staff, 2019, p. 2) While graduates may not immediately use the skills taught in certain modules, it is virtually certain that those graduates employ the foundations in all of the BSAP course modules at some point in their career.

GROUNDING IN PRACTICE

In contrast to professional military education programs that confer graduate degrees and must meet the accreditation standards for those degrees, BSAP is a school grounded in practice, taught by practitioners. Rather than a traditional staff college or war college, the demographics of BSAP students range from captains who have just completed company command to lieutenant colonels who are late transfers to FA 59. The civilian education of BSAP students is even more varied, with bachelor's degrees at one end and doctorates (typically in history or political science) at the other. Demographically, the FA 59 population is heavily weighted towards combat arms, most often infantry, armor, and field artillery, although the functional area has inducted officers from most of the Army Competitive Category (i.e., command track) branches.

As a common benchmark for assessment, all prospective BSAP students must complete a Graduate Skills Diagnostic (GSD) that assesses existing facility with the English language in a graduate education context. Taken without notes or study aids, the GSD examines three domains: (1) the structure of American English (grammar), (2) general language facility, including punctuation and mechanics, and (3) fundamental research protocol. Prospective BSAP students also must complete a diagnostic essay that assesses actual written products, rather than the automatically-graded assessments in the GSD. While neither is a screening mechanism, they do provide advance warning to the faculty of students who may be at risk of academic difficulties during the course (Applied Communications Lab, 2020).

These screening assessments ensure that only those officers who are critically interested in transferring to FA 59 actually compete, and that there is a clear quality cut that occurs prior to arrival to BSAP. The diagnostics prior to BSAP attendance provide some empirical evidence for predictive analysis of a student's expected performance during the course. While there are BSAP graduates who had mediocre diagnostic scores, they are often outliers. Finally, student performance in BSAP does not correlate with rank or prior experience in a basic branch.

Given the role of BSAP as a qualification course, foundations are even more important than they would be for more experienced officers studying those topics during a senior service college after battalion command. To deliver the BSAP curriculum to students who often have no prior experience in any of the functional area's knowledge, skills, and behaviors, the faculty are all experienced FA 59s with prior leadership experience in the functional area. The civilian faculty and director have served as division chiefs at the 4-star theater and national level and above, and the faculty can claim expertise across all of the FA 59 assignment categories and all of the echelons where FA 59 officers serve.

That basis of leadership experience and strategy expertise, much like the BSAP curriculum, provides a unity that is greater than the sum of its parts. Division chief experience is especially valuable because of the nature of work in FA 59 and the need to integrate skills across policy, strategy, and planning, both operational and institutional. The expertise across the faculty represents over 50 years of combined experience in uniform as FA 59 strategists at the division, corps, army service component command, joint task force, combatant command, army command, and at the national level, both service and joint.

As a corollary to that experience, BSAP grades its students in accordance with the same rubrics used at the US Army War College (Department of Military Strategy, Planning, and Operations, 2021). The enforcement of those standards is informed by the leadership experience of that faculty, who all served formerly as division chiefs at the joint task force, combatant command, or national level. The principal drivers of that enforcement are the expectations that BSAP graduates face daily in their organizations. Even in a three-star command, the standards are virtually indistinguishable from that of their 4-star higher headquarters. As a result, much of the BSAP curriculum teaches what is to be expected of a senior service college (apart from standards of grade), of which the following appears in Army Regulation 350-1:

> A military member O–5 and above, or Army Civilian GS–14 (or equivalent) and above, or who occupies a leadership position

(both command and staff) that requires a thorough knowledge of strategy and the art and science of developing and using elements of national power (diplomatic, informational, military and economic) during peace and war. This knowledge is necessary in order to perform Army, Joint, or Defense Agency operations at the strategic level (ACOM, ASCC, DRU, Field Operating Agency, Joint Task Force or higher) (Office of the Deputy Chief of Staff, G-3/5/7, 2018).

Most officers coming out of the battalion and brigade command track have few broadening opportunities above the tactical level unless nominatively selected for such opportunities; even then, those officers are handicapped by the requirement to command battalions and brigades in their basic branches. In comparison, a FA 59 officer will have amassed two to three times the amount of experience in strategy-related duties relative to command track officers by that point in their careers, because the FA 59 strategists will have first seen those duties as senior captains or as junior majors.

The correspondence of BSAP's curriculum to senior service college curricula makes it fundamentally different from other professional military education programs for officers of the same rank. One of the most common comparisons of BSAP is to advanced military studies programs (AMSP) such as the one administered by the School of Advanced Military Studies at Fort Leavenworth, Kansas or the School of Advanced Warfighting at Quantico, Virginia. While both share examinations of strategy and operational art, they differ in their grounding and focus of attention.

An AMSP makes tacticians into capable operational art practitioners. In comparison, BSAP, in enabling its graduates to critically assess and creatively develop strategic plans and policy, deliberately breaks its graduates of inappropriate tactical bias.

Instead, graduates of BSAP are trained to look from policy to strategy with operational art as a method of strategy implementation. Whereas AMSPs examine operational art from the bottom upwards as a nexus between tactics and strategy, BSAP examines operational art from the top downwards, owing to its focus on strategy as a bridge between operational art and policy. Instead of the division or corps where AMSP graduates serve, BSAP graduates are expected to be comfortable working at the combatant command or national level as a matter of course (Shekleton, 2014). While strategists (with occasional exceptions in the Army National Guard) do not command after entering FA 59, they trade in that obligation for command for deep expertise in the roles, missions, and functions of the Department of Defense at the national, combatant command, and theater levels. As a result, experienced strategists develop and exercise skills in policy, strategy, and operational art that no other branch or functional area can match.

What BSAP Produces

By the end of their program of instruction, BSAP graduates are prepared to serve as Army strategists in Army, joint, or multinational organizations. As a function of the instruction in the course, they will possess a rich professional perspective on policy, strategy, and doctrine. Six program learning outcomes represent the expected capabilities of a graduate:

1. Synthesize and evaluate existing strategic and operational theory as an explanatory framework for critical analysis of the practice of strategy, the creation and application of doctrine, and the design of future strategies.
2. Drawing upon a synthesis of strategic theories and historical insights, evaluate grand, national, and military strategy in selected case studies.

3. Evaluate today's complex international system and domestic context, competing approaches to national security, and the national security decision-making process.
4. Create military options based on an evaluation of US regional interests and objectives, trends, and theater strategic factors in regions of strategic interest to the Joint Force.
5. Evaluate the systems that manage, develop, and transform the Army and Joint Force.
6. Understand the Joint Strategic Planning System and the relationships between theater strategies, combatant command campaign plans, contingency plans, and crisis action plans (Department of Military Strategy, Planning, and Operations, 2022).

Graduates of BSAP can apply these outcomes to any type of assignment, whether operational, institutional, or applied. Building on top of the intellectual foundations in civilian education and practical foundations in professional military education, they will be prepared for continued self-development and eventual mastery of strategist competencies throughout the duration of their careers. In the nineteen years since the start of the BSAP program, the myriad contributions of its graduates have ranged from joint task force operations orders to combatant command war plans to national strategy documents at the highest levels. Future graduates of the program can and should expect to build upon those achievements. [See Appendix for corresponding PowerPoint presentations.]

REFERENCES

Applied Communications Lab. (2020). *Communicative Arts Directive, Distance Education Program Class of 2021*. U.S. Army War College, Strategic Studies Institute. https://ssl.armywarcollege.

edu/dde/DDE_documents/DEP_CAD_AY21.pdf

Chairman of the Joint Chiefs of Staff. (2019). *Description of the 2018 National Military Strategy of the United States of America*. Joint Chiefs of Staff.

Chairman of the Joint Chiefs of Staff Instruction 3100.01E, Joint Strategic Planning System. (2021). Joint Chiefs of Staff.

Chilcoat, R. A. (1995). *Strategic Art: The New Discipline for 21st Century Leaders*. Strategic Studies Institute, U.S. Army War College.

Department of Distance Education. (2020). *Memorandum for Battalion and Brigade Commanders of Command and General Staff Officer Course Distance Learning Intermediate Level Education Students, Subject: Support of Students Completing CGSOC via Distance Learning*. U.S. Army Command and General Staff College.

Department of Military Strategy, Planning, and Operations. (2021). *Basic Strategic Art Program Evaluation Guide*. U.S. Army War College.

Department of Military Strategy, Planning, and Operations. (2022). *Basic Strategic Art Program AY 2022(B) Course Introduction, Strategic Theory Directive, Strategic Art Directive*. U.S. Army War College.

Galvin, J. R. (1989). What's The Matter With Being A Strategist? *Parameters, 19*(1), 2–10. https://doi.org/10.55540/0031-1723.1522

Headquarters, Department of the Army. (1999). *Field Manual 22-100, Army Leadership*. GPO.

MacMullen, R. M. (2009). *Regular Army FA59 Strategic Plans and Policy Update*. Headquarters, Department of the Army.

Moore, C. P. (2009). What's the Matter with Being a Strategist (Now)? *Parameters, 39*(4), 5–19. https://doi.org/10.55540/0031-1723.2500

Office of the Deputy Chief of Staff, G-3/5/7. (2017). *DA Pam 600-3 Smartbook, Strategist Functional Area*. Department of the Army.

Office of the Deputy Chief of Staff, G-3/5/7. (2018). *Army Regulation 350-1, Army Training and Leader Development.* Headquarters, Department of the Army. https://armypubs.army.mil/epubs/DR_pubs/DR_a/pdf/web/ARN18487_R350_1_Admin_FINAL.pdf

Office of the Deputy Chief of Staff, G-3/5/7. (2020). *DA Pam 600-3 Smartbook, Strategist Functional Area.* Department of the Army.

Officer Personnel Management System XXI Task Force. (1997). *Officer Personnel Management System (OPMS) XXI Final Report.* Headquarters, Department of the Army.

Park, F. J. H. (2015). A Framework for Developing Military Strategists. *Infinity Journal, 5*(1), 9–14.

Shekleton, M. A. (2014). *A Comparative Analysis of the Basic Strategic Art Program (BSAP) and the School of Advanced Military Studies' (SAMS) Advanced Military Studies Program (AMSP).* Center for Strategic Leadership, U.S. Army War College.

U.S. Army Command and General Staff College. (1977). *College Catalog, 1977-1978* (College Catalogs). Combined Arms Research Library.

U.S. Army Human Resources Command. (2010, December 20). U.S. Army Human Resources Command examines Voluntary Transfer Incentive Program (VTIP), among other. *ArmyNews.* https://www.army.mil/article/49699/u_s_army_human_resources_command_examines_voluntary_transfer_incentive_program_vtip_among_other

9

NATIONAL SECURITY AND THE HISTORIAN'S ETHOS

Anthony Eames

In his first press conference after being announced as the next president of Columbia University in the summer of 1947, General Dwight Eisenhower promised "Wherever I am...national security will always be my number 1."[1] Eisenhower followed through on that commitment, taking leave to serve as NATO's first Supreme Allied Commander in Europe in 1951 and 1952. At the university, Eisenhower's primary interest was to secure funding for and direct a massive research project on *The Ineffective Soldier*.[2] Despite Eisenhower's heralded administrative skills, Columbia did not thrive under his leadership because his obligations to the university were subservient to his obligations to national security. This model has all too often been considered by national security leaders to arrange a military on top-higher education on tap relationship.

General H.R. McMaster's recent essay, "Preserving the Warrior Ethos," in the *National Review*, argues for exactly such an arrangement.[3] The warrior ethos, as McMaster understands it, has been a historical constant since the days of Achilles. American's social, political, and moral values, however, are more dynamic than ever before. The dissonance between these two phenomena erodes

---

1 "Security is First, Eisenhower Says," *The New York Times*, 28 June 1947.

2 See Eli Ginzberg oral history at Dwight Eisenhower Presidential Library. https://www.eisenhowerlibrary.gov/sites/default/files/research/oral-histories/oral-history-transcripts/ginzberg-eli.pdf

3 H.R. McMaster, "Preserving the Warrior's Ethos," *National Review*, 28 October 2021.

the will of American soldiers to fight and win wars. McMaster identifies the study of history as the antidote, specifically military and diplomatic history rooted in the great man tradition.

However, General McMaster's recommendations for preserving the warriors ethos are untenable because he fails to understand the historian's ethos. McMaster would do well to recall the work of one of his favorite historians, Marc Bloch. Bloch wrote his masterpiece, *The Historian's Craft*, while serving in the French resistance in the Second World War. In his work, Bloch made the case to "preserve the broadest interpretation of history."[4] The form of history that Bloch advocated for—the *Annales school*—stressed perspectives from ordinary people. This understanding of history underwrote Bloch's profound sense of duty to fight for France in both World War I and World War II.

Bloch's belief in studying the perspectives of those dispossessed of power and writing history based on the widest variety of evidence laid the foreground for the New Left's writing of history from the bottom up and the socio-cultural-environmental turn among professional historians that McMaster holds responsible for sewing the divisions in American society that undermines soldiers will to fight. But one must ask if these now paradigmatic features of the historian's ethos are really an anathema to national security. Should historians distort their craft to preserve the warrior's ethos?

The answer of course is no, the US military and defense community should embrace the historian's ethos to better craft national security priorities reflective of the whole of American society. Eliot Cohen, certainly no stranger to the academy and the armed services, cautioned that military professionals are "more likely to misuse history" for their own immediate ends rather than invest— as he recommended—in the acquisition of the "historical mind."[5]

---

4   Marc Bloch, *The Historian's Craft* (New York: Vintage, 1964).

5   Eliot Cohen, "The Historical Mind and Military Strategy," (Fall 2005): 575.

The Meaning of National Security

Ironically, the very idea of national security is a product of historians grappling with the best way to make sense of the purpose and sources of American power while also coming to terms with an evolution in their professional methodology. The concept of national security has its origins in the progressive era. Albert Bushnell Hart—known as "The Grand Old Man of American History"—was among the first to advance the term "national security" as a leading member of the "National Security League" that rose to prominence during and immediately after World War I.[6] Hart considered his calling to be the professionalization of history in higher education. As a founding member of the *American Historical Review*, the journal of record for the American historical profession, Hart advocated for "scientific history." His definition of the discipline emphasized baseline standards of peer-review, research, and employment of evidence to help cure the field of the plague of forgeries and unfounded assertions that had made history into a handmaiden of politics throughout the 19th century.[7]

Hart's historical ethos underpinned his service to HBCUs, and other racially inclusive causes that steeled him in his fight against the fascist impulses of the National Security League. The most prominent fight being over the organization's promotion of the 100% Americanism principle, which sought to reign in the freedoms of America's enemies at home, which included foreign nationals, labor unions, pacificists, members of Congress opposed to the League's agenda, and bizarrely—Wisconsinites. The League's failure to heed Hart's warning that the divisive 100% Americanism principle was itself a threat to national security was a major cause of its decline. Although many of the League's policies for preparedness would come to pass, its divisive approach is one

---

6   Mark R. Shulman, "The Progressive Era Origins of the National Security Act," *Dickinson Law Review* (2000): 289-330.

7   Hart, imagination in history…the Hart obituary, nat sec article, etc.

reason the phrase of national security failed to take hold in the American lexicon during the interwar years.[8]

Amidst rumblings of the coming War, historians revived the term national security as a new code of professional methodology and obligation manifested in their discipline. Edward Meade Earle founded the Princeton Study Group in the 1930s to study problems facing the international community and their impact on American society. Within his profession, Earle identified with a generation of historians who rejected "scientific history," in favor of what they referred to as "new history."[9] Disciples of new history believed the present to be germane to the study of the past. Whereas scientific historians tended to focus on official records of public institutions in constructing narratives of high politics and diplomacy for an erudite few, practitioners of new history believed that uncovering the private lives of our ancestors—both powerful and powerless—offered an important service to contemporary society.[10]

The new history approach to studying the past from the bottom up, and middle out, gave the Princeton Study Group a model for how America should engage the world for the benefit of the widest swath of American society. Despite the earlier efforts of the National Security League, in the 1930s and early 1940s the terms national defense and national interest were the most prominent phrases in discussions that we would later consider under the umbrella of national security. It was Earle's group at Princeton whose approach to scholarship gave them a more inclusive view of the foundation of American values that ultimately entrenched the term national security in government and in higher education. They stitched together the meaning of security as understood in the context of

---

8   Andrew Preston, "Monsters Everywhere: A Genealogy of National Security," *Diplomatic History* (June 2014): 487.

9   David Ekbladh, "Present at the Creation: Edward Mead Earle," *International Security* (Winter 2011/2012): 113.

10   James Harvey Robinson, "The New History," *Proceedings of the American Philosophical Society* (May-August 1911).

the New Deal social security program and the Wilsonian idea of "collective security" with the preference for national action found in the terms national defense and national interests. This hybrid term represented a compromise between American individualism and American internationalism that helped to marginalize the remnants of isolationism in the United States, which had been established on the 19th century idea of "free security" and the early 20th century notion of "continentalism."[11]

Current historians should take note of the ways the Princeton Group engaged both the government and critics of government. It was the Princeton Group that schooled the "father of American journalism," Walter Lippman, in the concept of national security. In his 1943 best-selling volume *US Foreign Policy: Shield of the Republic*, Lippman deployed the concept of national security as his "controlling principle" to keep US commitments abroad in balance with US resources and US values.[12] Similarly the Princeton Group advised James Forrestal as he formulated and lobbied for the legislation of the 1947 National Security Act that would elevate him from Secretary of the Navy to the first US Secretary of Defense.

The key innovation of the phrase national security was that it united the long-held emphasis of territorial defense with the novel sense of the need to defend ideology.[13] Already by 1952, it was this latter aspect of national security that the Yale historian and political scientist, Arnold Wolfers, held responsible for making the term into an "ambiguous symbol."[14] National security presented a "moral problem" for policymakers who first had to determine what values defined American ideology in a morally heterogenous society and

---

11 Emily Rosenberg, "The Cold War and the Discourse of National Security," *Diplomatic History* (1993): 278-279.

12 Walter Lippmann, *U.S. Foreign Policy: Shield of the Republic* (New York: Little Brown Books, 1943).

13 Preston, "Monsters Everywhere"

14 Arnold Wolfers, "'National Security' as an Ambiguous Symbol," *Political Science Quarterly* (December 1952).

deserved protection beyond water's edge. Furthermore, Wolfers articulated two more ambiguous qualities to national security. First, every increment of security had to be paid for by additional sacrifices of other values. Writing at the peak of McCarthyism, Wolfers' point hardly needed illustration. Second, the overzealous pursuit of national security ran the risk of creating a security dilemma. Maximizing one's security without prompting a response from an adversary that in turn reduced one's security was itself an ambiguous threshold based on internal and external threat perceptions.

An Expanded Role for Historians

By 1980, the ambiguities of national security had been excessively poured over by historians to explain the cause and consequences of the Cold War. Orthodox accounts and revisionists explanations battled over whether the correlation of forces, the spread of ideology, or economic imperatives animated US foreign policy and whether that policy was imbued with imperial ambition or a commitment to containment. Thirty years of insular debate along these lines led prominent Harvard historian Charles Maier to declare that diplomatic history had become a "stepchild" of the historical profession. Yet another thirty years after Maier's lament, diplomatic historians celebrated the fact their field had become a "bandwagon" of the broader profession. So what changed? The answer is that the New Left and their academic offspring revitalized the study of US foreign relations by bringing forth an entirely new set of categories for analysis that were not and still largely are not featured in discussions about national security; namely: race, class, gender, and environment.[15] In the last thirty years, the framing of diplomatic and military history has also shifted from the national perspective to the transnational view.

General McMaster is suspicious of historians who emphasize

---

15 See the Diplomatic History Roundtable in *Journal of American History*, 95 (2009).

these categories in their analysis, and he is not alone given that the ratio of liberal to conservative college and university faculty has doubled from 2.3:1 in 1989 to 5:1 in 2016.[16] In conjunction with this trend, trust in higher education has fallen dramatically. In 2019, only 50% of Americans believed that colleges and universities had a positive impact on American society.[17] Declining trust is also a significant problem for the US military. According to the 2021 Reagan National Defense Survey, only 45% of Americans expressed a great deal of trust and confidence in their military, which is 25% drop since 2018. Unlike higher education, the decline in Americans' view of the US military is bipartisan. Republicans rate of disapproval over the past three years has metastasized at nearly twice the rate of Democrats, though overall Republicans still voice a higher approval of the military than Democrats.[18] Also unlike higher education, the US military trends more conservative. Reinforcing the positive relationship between these two institutions may go a long way in ameliorating the harmful effects of political polarization, improve civil-military relations, and ultimately benefit US national security.

One way to realize these goals is to recognize that in historical scholarship—as in the practice of medicine—the diagnosis may be correct, but there can be disagreement on the course of treatment. Historians who oppose an assertive military posture are still capable of producing scholarship that provides an analytical benefit to the national security community. Mary Kaldor is a clear example of this. As one of the founders of the "peace studies" curriculum that took hold of higher education in the 1980s, Kaldor was staunchly opposed to increased investment in military capabilities. Peace studies was founded to present alternative solutions to

---

16   Rikki Sargent, Shannon Houck and Lucian Gideon Conway, "How to Stop Political Division from Eroding Military-Academic Relations," *Defense One*, 8 July 2021.

17   Kim Parker, "The Growing Partisan Divide in View of Higher Education," *Pew Research Center* (August 2019).

18   Ronald Reagan Institute, "Reagan National Defense Forum," (November 2021). https://www.reaganfoundation.org/media/358085/rndf_survey_booklet.pdf

international conflict than those advanced by security and strategic studies curriculums that sprang up in connection with the phrase national security. Edward Mead Earle, to no surprise, is the godfather of strategic and security studies.[19] Despite their opposite political agendas, security studies scholars learned a great deal from Kaldor's concept of the "baroque arsenal," a pattern in which military armaments have become decadent in their complexity and cost but may not be considered advanced because they are either ineffective operationally or fail to provide a new capability that changes the correlation of forces. Though 40 years old, Kaldor's concerns about "'trend innovation'" – perpetual improvements to weapons that fall within the established traditions of the armed services and the armorers," very much apply to the F-35 project and plans to develop a next generation W93 warhead for Columbia-Class nuclear submarines.[20]

Historians who have embraced the social-cultural-environmental-transnational approach have unveiled new dimensions to national security issues. Gender is, and has been, a major determinant in international conflict. Leading scholarship on the Spanish-American War demonstrates that it was largely driven by a postbellum crisis in American manhood perpetrated by yellow journalism.[21] The National Security League's premise of 100% Americanism, argued that Americanism meant "manhood."[22] Anyone who doubts the importance of gender analysis to national security issues would do well to do a Google search for one of the many shirtless photos of Vladimir Putin to better understand how the role masculinity in the legitimation of power in Russian society leads to a confrontational Russia.

---

19   Ekbladh, "Present at the Creation," 108.

20   Mary Kaldor, *The Baroque Arsenal* (New York: Hill and Wang, 1981).

21   Kristine Hoganson, *Fighting for American Manhood: How Gender Politics Provoked the Spanish-American and Philippine-American Wars* (New Haven, CT: Yale University Press, 2000).

22   Preston, "Monsters Everywhere," 486.

Race matters as well. The Nazi's dehumanization of the Jewish people and the US government's dehumanization of the Japanese were both important components of their wartime policies. Historians have shown that positive developments in US civil rights in the 1960s were directly related to the need to improve US foreign relations with the Third World so as to win over hearts and minds in the Cold War.[23] The sad reality that ethnic cleansing persists in so many different regions of the world today drives home the point that race analysis is critical for understanding national security challenges.

In the American context, New Left historians have shown that the deep interconnection between race and class have led to national security crises. One of the most striking examples of this phenomenon comes to us from Eric Foner, whose seminal study of the causes of the Civil War, revealed how a clash of economic philosophies and glorification of the middle-class in the United States produced the most significant national security crisis in American history.[24] The continuing importance of class analysis to national security is illustrated by the way President Trump's "America First" policy and President Biden's pledge to conduct foreign policy for the middle class led directly to the U.S. withdrawal from Afghanistan.

These advances that have returned diplomatic and military historians to respectability within the historical profession do not seem to have been incorporated into the ways the U.S. military's deploys history to better understand national security. The Naval War College's graduate level courses on national security decision making and theater security decision making do not include any significant treatment of gender, race, class, or even environment.[25]

---

23 Mary Dudziak, *Cold War Civil Rights: Race and the Image of American Democracy* (Princeton, NJ: Princeton University Press, 2011)

24 Eric Foner, *Free Soil, Free Labor, free Men: The Ideology of the Republican Party Before the Civil War* (Oxford: Oxford University Press, 1995).

25 https://usnwc.edu/Faculty-and-Departments/Academic-Departments/National-

The same is true for similar courses taught at the Army War College and the Command and General Staff College. The *National Security Primer* that is meant to provide National War College students at the National Defense University a common foundation on national security thinking is entirely neglectful of the type of analysis developed by historians following the social-cultural-environmental turn.[26] Instead, the likes of Thucydides, Sun Tzu, Clausewitz, Brodie, Mahan, and Schelling still reign supreme on syllabi and materials designed to teach national security.

Another issue is that long tenures have become the norm in the historical office at OSD, with only three scholars filling the post of chief historian for 72 of its 73 years in existence. With some notable exceptions, this is norm throughout similar historical offices in government. Does this produce the best possible historical outcomes; or does it lead to a historian becoming more attuned to the organizational culture of the DOD rather than evolving ethos of the historical profession?[27] The United Kingdom may provide a better arrangement, in which historians are seconded from their university positions. The official historians of the British nuclear deterrent in particular have produced excellent volumes that at the time of production brought to bear the latest developments in the history of science and technology, environmental history, and security studies.

One impediment to this tasking model is the strict secrecy provisions preventing historians from gaining access to the sources they need to conduct a thorough accounting of national security decision-making in government. Those who are suspicious of the social and cultural explanations of national security must understand that the dramatic increase in classification over the

---

Security-Affairs-Department

26 https://nwc.ndu.edu/Portals/71/Documents/Publications/NWC-Primer-FINAL_for%20Web.pdf?ver=HOH30gam-KOdUOM2RFoHRA%3d%3d

27 https://history.defense.gov/Historical-Office/Past-Chief-Historians/

past forty years has also left historians searching for new sources. In 2015, classification decisions occurred at nearly 140 times more often than they did in 1995.[28] A recent look at military-academic relations has revealed that equality between partners is crucial for building trust. While that study dealt with issues of academic and military rank, it should also be recognized that inequality in regard to access to information generates mistrust and missed opportunities.

Related to this issues is that the I in the DIME construct of national security tools has been undervalued compared to D, M, and E. Putin's attempt to rewrite the history of Ukraine to justify his war is just one instances in which historians can and should be employed as "information warriors." In a tight job market, in which only about 50% of history Ph.D.s land academic jobs, well-trained historians are on the hunt for opportunities. All too often, the Russian nuclear historian, the scholar of Latin American drug cartels, the Iranian oil historian, the historian of Sino-Japanese economic relations, or the North African environmental historian forsake their opportunity to work for the intelligence community, the State Department, and the DOD. Their time in country working in faraway archives and within local communities makes a long security clearance process still longer and they cannot afford to wait. There are too few pathways for trained historians to acquire a security clearance, with the best and sometimes only options for obtaining a clearance being through employment with a defense contractor. Universities, perpetually claiming to be short on funds, are loathed to sponsor a security clearance for graduate students or even tenured faculty. The secrecy regime in effect seals off a potentially huge pool of information warriors from aiding national security efforts.

---

28   James Bruce, Sina Beaghley, George Jameson, "Secrecy in U.S. National Security: Why a Paradigm Shift is Needed," RAND Report (November 2018),14.

Pondering A New Paradigm

Is it time for historians to initiate a reconsideration of the term national security all together at a time when the sources and purpose of American power appears out of sync? "National sustainability" represents an intriguing evolution of the phrase national security that historians and military professionals would do well to ponder. "Sustainability carries its own political baggage because of its association with the progressive politics that McMaster holds responsible for the erosion of American security, but the overt environmental element could bring about a new alignment of political support for the development of US power and its applications in competition with great power competitors and global threats. It may appeal to Republicans between the ages of 18 and 39 who are twice as likely as their conservative elders to consider climate and environmental issues as a series problem.[29] The phrase national sustainability also brings to mind the issue of climate migration, which will continue to exacerbate border security, an issue that Republicans identify as a national security priority, but not Democrats. National sustainability holds the promise of mobilizing new minds to resolve the liabilities of the US military's commitment to a baroque basing model that has left dozens of military installations around the world highly vulnerable to rising sees, severe weather, and extreme temperatures.[30] National sustainability serves as a call to action for the United States to master the strategic energy pivot from hydrocarbons to green energy technologies. Lastly, national sustainability may lead to a whole of government and whole of society approach to better tailor mission to capability and facilitates an equipoise between resources and commitments.

---

29   Jeff Brady, "Light Years Ahead of the Elders, Young Republicans Push GOP on Climate Change," *NPR*, 25 September 2020.

30   https://www.americansecurityproject.org/climate-energy-and-security/climate-change/climate-change-and-u-s-military-basing/

10

ENCOURAGING DUAL-ENROLLED STUDENTS TO ENROLL IN CORPS OF CADETS AT SENIOR MILITARY COLLEGES: BARRIERS AND OPPORTUNITIES

Panelists:
- Imani Cabell, Ph.D.
- Katherine Rose Adams, Ph.D.

As presented at the 2022 Strategic and Security Studies Symposium
Hosted by the Institute for Leadership and Strategic Studies
University of North Georgia

Dr. Keith Antonia:

Imani Cabell is the assistant director of dual enrollment at UNG and a doctoral student in the UNG Higher Education Leadership and Practice, or Ed.D., program. Imani's passion centers on eliminating student barriers to a college education and promoting impactful student programs. Her experience working with a variety of students from different high schools and walks of life has enhanced her understanding of program specifications and potential barriers that may impact student success. As a doctoral student, Imani's knowledge and support for student achievement have been enhanced through research on specialized student populations and alternative college admissions opportunities. Through years of working in higher education, Imani is still excited by the opportunity to help students achieve their academic goals, value their education, and progress toward a successful future.

Dr. Katherine Rose Adams is the assistant professor for, and the program coordinator of, the Higher Education Leadership and Practice doctoral program here at the UNG. Katherine teaches coursework on higher education leadership theory, qualitative research, student affairs administration, and law and ethics in higher education. Katherine's research interests are in the areas of boundary spanning, community engagement, higher education trends, university-community partnerships, collegiate leadership, and research communication. Katherine received her Ph.D. in adult education where her focus was on the roles, characteristics, and motivations of community leaders' boundary spanners within the university community partnerships. She also obtained a Master of Education in human resources and occupational development and has a master's certificate in interdisciplinary qualitative research and a bachelor's degree in psychology from the University of Georgia. She is an associate editor for the *Journal of Community Engagement and Scholarship*. Dr. Cabell and Dr. Adams will now make their presentation. Thank you very much.

Dr. Katherine Rose Adams:

Thank you all so very much for being here today. We're definitely taking a lot more of a higher education lens with our presentation and what we're going to look at is this potential for a new population. We're going to present today on encouraging dual enrolled students to enroll in the corps of cadets at senior military colleges, and we're going to address some barriers and some opportunities for that.

The why this is important is that enrollments in US colleges and universities have fallen for 10 years consecutively, and the impacts of Covid 19 have only exacerbated this for recruiting and retaining students. Since the outbreak of Covid 19, enrollment losses and the pandemic represent a two-year total of 5.1 percent, but what this means overall nationally is about 940,000 students. So far, we're seeing a reduction in the last two years; also, Generation Z, has begun to reveal a hesitancy over the hidden and real costs of higher education. We may see this yield a significant decline in the coming decades, leading to even tighter competition for students. The forecasting of an enrollment cliff in higher education can quickly turn into a significant fiscal crisis seen with the consolidations of higher education institutions as well as closings. To combat the predictions of tighter resources, declining enrollment, and changes in a generational perspective, it is time to seek recruitment efforts toward new and untapped applicant populations.

I don't think, if we were doing one of the educational programs, we would have to tell you what the corps of cadets is, but for this instance, we're going to situate how we are presenting this information within the state of Georgia and utilizing UNG as our case example. So a little bit about UNG's corps of cadets. We're one of six senior military institutions. UNG was founded in 1873, the same time as the corps began. UNG is also the first institution to have women enroll. It was also 20 years prior to any other senior military institution that had a female cadet. UNG corps of cadets averaged about 721 cadets annually (we average about 750 this last year) and is consistently ranked as the nation's best army ROTC program.

Dr. Imani Cabell:

Now I'm going to talk a little bit about what is dual enrollment and what is dual enrollment in this landscape. First and foremost, dual enrollment is a unique opportunity for high school juniors and seniors, rising juniors and seniors, to take college-level courses while simultaneously completing the necessary classes that they still have in their junior and senior year to reach the end of their high school level diplomas. With this opportunity, dual enrolled students can both engage in high school level coursework and still be in the members of their organization while completing their college-level core curriculum courses.

Let me give you a little bit of the benefits of why the dual enrollment program is so beneficial especially here at UNG. First and foremost, the University System of Georgia and Georgia as a whole has bought into dual enrollment. Dual enrollment is a free program for students. Students can take up to 30 credit hours and have that covered by the state of Georgia. Each student does have a maximum of 15 credit hours that they can take per semester, but 30 hours will max out their allotted funding within the program. We know that a lot of our students here at UNG and within the university system tend to take advantage of our Zell scholarship and Hope scholarship, which is our lottery-based funding of aid. So dual enrollment programs do not infringe on the 127 hours of Hope and Zell eligibility.

To give you a tangible example, I typically have seniors in high school have the opportunity to finish their full freshman year before they graduate high school. UNG has one of the largest dual enrollment programs. I've had roughly about 37 high school graduates finish our program and gain their associate's degree prior to graduating high school.

What does that mean for UNG as a whole? What does that mean for this conversation? Students are getting an opportunity to get a jump start on college, the full immersion of college opportunities—engaging in scholarship opportunities, leadership classes, things of that nature—students can be fully immersed at the university level, but they can still engage in their high school. They can go to prom; they can be involved. They can still have that high school experience but gain access to academia, gain the college perspective as well.

Let's talk a little bit about course modalities. Students can take a variety of modalities. They can take their classes fully online, they can take them in a hybrid fashion, or they can take them fully in person on campus. So it's really about the nature of that student's experience. What they want to gain and how they want to go about doing so.

To give a brief history on UNG's corps of cadets and dual enrollment, back in fall 2013, UNG consolidated. We had dual enrollment on all four campuses at that time. We had roughly about 250 dual-enrolled students. Now our program here at UNG has seen immense growth. Just by that graphic right there, you all will be able to see we started at about 250 dual enrolled students. We have now topped out upwards of 1,571 dual enrolled students. What does that mean? These students are going to be some of your best and brightest. They are going to be the ones who are interested in their education. They're going to be interested in the opportunities that college can present to them. They're going to want to be your leaders. One thing that we have learned throughout this symposium is some of the corps of cadets' great values of developing and enhancing leadership. These students are coming in, and they're wanting to grow, they're wanting to be present, and they're wanting to do activities where they have to be present-minded.

By utilizing the UNG dual enrollment program in conjunction with the corps of cadets, we viewed this as a great opportunity to engage a different level of thinker when it comes to students. We have had some dual enrolled students here at UNG participate in the corps of cadets. As you can see, we haven't had many. Roughly, about a total of seven applicants over our span of time (that does not include students who brought completed credit in that span of time) did dual enrollment at another institution; this is more so geared towards students who participated in UNG's dual enrollment program. But we have had roughly about seven applicants. One who attended and graduated as a UNG cadet and of whom Dr. Keith Antonia is very proud. We've had four students at this time who have had canceled applications, so that maybe opened up an opportunity to figure out why those applications have been canceled, and two students who were denied. The reason we have such a growing dual enrollment population is that we hold our students to a standard. We want to make sure they're not only academically ready but also mentally ready and emotionally ready to be in college-level classes, so they have to meet a higher standard than our traditional UNG freshman student.

Some potential barriers to success that may come up with bringing dual enrolled students within the corps of cadets is that, typically, UNG corps of cadet students are required to live on campus. When you think about a high school student, they must be 17 at the start of the semester to participate in the corps of cadets. Trying to figure out how to navigate that age as well as the living requirement can be a barrier to success. Balancing a lot of responsibilities: one thing as a dual enrollment advisor that I've worked with many students on is trying to navigate your responsibilities. How do I not only stay ahead in my classes but also stay on top of my other responsibilities within my athletics or just the organizations in which I'm participating? How do you navigate that balance, getting students to not only navigate that but also prioritize their academics? We always want to talk about advising from our perspective, as advising is teaching, so we always want to ensure that they know how to not only do well in the program but also continue to thrive as students. Then we also have to make sure we have college readiness. Some of these classes are going to devote more rigor than some of the traditional high school courses, so they have to make sure they have that college rigor, that level of maturity, and they really can handle the college-level coursework.

One additional barrier to success is UNG's frog week for cadets. Typically, that frog week happens on the first week of high school, so that can be a barrier for some of our local high school partners. Trying to figure out the best way to navigate that barrier so students can still participate in the activities that happen during frog week but still go to the first week of school.

The last barrier to success would be making sure that classes that fall under the military science requirements can get dual enrollment funding. Part of what we do as dual enrollment and as our dual enrollment team here at UNG is that we go through and classify each class so that it can gain state funding.

Dr. Katherine Rose Adams:

One of the reasons we want to make sure that we are covering those barriers is that we can look at those and come up with innovative solutions on how we might work around them or how might we find ways to have partnerships. So we have some recommendations for what we see potentially being success of this population. The first would be to strategically recruit from high schools with either greater flexibility or with a common mission. Our one graduate that we had highlighted earlier was someone who attended Georgia virtual schools. As high school students attending fully virtual, this gave them different flexibility in being able to navigate the space and being able to be on campus for all the responsibilities of being a cadet.

The second piece would be to create an admission advisor who's designated for this population in order to support the application material. As we saw, four of seven of the applications weren't completed. That's because there are multiple levels to those applications. There's the admission, there's the dual enrollment, there are the responsibilities, and the responsibilities and requirements of applying to be a cadet. Having someone to help support and sort of navigate that process might be beneficial.

The other piece would be either a split option or a summer option. A split training option would allow 17-year-old juniors in high school to join the army reserves or army national guard training as a senior with a local unit one weekend per month. Or the idea of a summer enrollment option would be allowing the summer beforehand, before senior year to attend a tech school to get some of their military science credits that would not have been covered under dual enrollment.

The next piece would be thinking about recruiting in different fashions. Whether that would be by extracurriculars, such as UNG's Olympic producing rifle athletic team, or if this was something that other high schools had as a way to sort of gatekeep and come in or to go and really look at programs that have strong junior ROTC programs.

Another piece, and this I think is one of the most valuable pieces that we've just recently learned, is with the approval of military science coursework; right now, our state governing board, the University System of Georgia, currently has approved 28 military science courses. However, at this time no institution in the state has gone and begun developing their coursework to be approved through Georgia tracks, which is the system that works within dual enrollment. That would be something to go and begin to create potentially a military science dual enrollment pathway program which would help us lead into this area.

And then the last, as Imani mentioned before, the inclusion of Hope and Miller funding would also allow for the cadets and students, once they're fully enrolled, to use their Hope hours to extend their time. That could be a double major; it could be going and creating minors; it could be some of the new four plus one programs where if you still hold on to one of your undergraduate credits, you could begin taking graduate credit and still have your Hope hours covering that.

These were some of the recommendations that we were seeing to kind of address some of the barriers we previously talked about.

Dr. Imani Cabell:

Through this presentation, through just some of our great collaboration and partnerships that we have here at UNG, we wanted to look at some of the first steps. In speaking to some of the military leadership in this room, such as Dr. Keith Antonia, we are now looking at taking some of these first steps to just see about the possibilities of integrating a relationship between dual enrollment and UNG's corps of cadets. We're looking at the possibility of integrating some cadet materials and marketing materials within our dual enrollment admissions groups. For example, we tend to go to roughly about 30 to 35 high schools each semester, so just having some of this military and core cadet information with us gives us an opportunity to bridge that proverbial gap between high school and college transition and just start to introduce some of the bigger benefits of the corps of cadets to these potential dual enrolled students. To give you guys a brief example, Dr. Antonia provided me with some material these last two weeks. I might have been in about five high schools, and I am now out of material— you need some more material—but our students are wanting to engage, they are wanting to learn, and just if we target our recruitment efforts together, if things like UNG's advising is going into schools, then also UNG's corps of cadets can go in, and now we are partnered together to really get a hand on some of that recruitment. And then as Dr. Adams said, there is a very big opportunity for us to get some of these military science classes approved through the University System of Georgia and Georgia tracks. There are 28 approved courses currently on the TCSG side, so that is a very easy process for us to try to go about integrating some of that into the USG side just so cadets and future cadets and potential cadets can have an additional opportunity.

Now part of us being higher education professionals is that military leadership is not necessarily our frame of space, but one thing we think is very valuable is to continue having conversations about how to navigate this new version of higher education that we are all experiencing. As Dr. Adams said during one of her slides, one of the big things that students are seeing is a move and a transition to online learning. So, how can we incorporate this new generation of thinkers, this new more hands-on online learner, and then bring them into dual enrollment, bring them into the corps of cadets, and bring them here to UNG so that they can gain some of these resources, so that we can speak to them, and they can benefit from the things that UNG has to offer. So we would like to thank you guys for allowing us to come and have a conversation in this space. Thank you for opening the doors to have some of these conversations with us.

11

HIGHER EDUCATION AND NATIONAL SECURITY IN THE USA

Cadet Natali Gvalia

As presented at the 2022 Strategic and Security Studies Symposium
Hosted by the Institute for Leadership and Strategic Studies
University of North Georgia

Cadet Natali Gvalia was born in 2002 in Tsunaki, Georgia. She has two brothers and a sister. In 2008, she attended public school in Tbilisi and received an academic achievement award: the golden medal. She has volunteered for workshops marathons and wrestling tournaments, and, after the 12th grade, she passed national exams to get enrolled into the national defense academy in Gory where she successfully completed basic combat training and became a first course younger performing well both physically and academically. She's now in her second year of studies and hopes to work as an officer in the near future.

Cadet Natali Gvalia:

I'm Natali Gvalia from the National Defense Academy of Georgia, the Republic of Georgia, which is in the middle of Europe and Asia. I'm going to give you brief information about US national security and strategy. I would like to start with a little bit of background information. The great struggles of the 20th century between liberty and totalitarianism ended with a decisive victory for the forces of freedom and a single sustainable model for national success, freedom, democracy, and free enterprise.

Today, the United States enjoys a position of unparalleled military strength and great economic and political influence. I cannot talk about US national security without emphasizing 911. To do so, I'm going to use President George Bush's quote:

> Terrorists attacked a symbol of American prosperity, but they didn't touch its source. America is successful because of the hard work, creativity, and the enterprise of our people.

I have the question, what are the things which those things come from? Like hard work, creativity, and enterprise of their people? What are these things? I have the answer, and this is education.

Education is the password to the future, and tomorrow belongs to those who prepare for it today. Welcome to my presentation about US national security and higher education. Let me start by giving you some background information. Here is the presentation plan you can see here. First, we are going to explore the US higher education system. Then we are going to talk about US national security strategy; US education reform and national security, a matter of national security and how K-12 education impacts America's military; and lastly, we are going to talk about what's America's biggest national security issue, and it's the K-12 education system, of course.

Let's explore the US higher education system. Every student deserves the chance to get high-quality education no matter where they live or how much their family makes. Strengthening the American education system from K-12 to higher education to career and technical education is good for students, communities, and the economy. Firstly, the American public system and education is in need of comprehensive reform to ensure that all students are able to realize their full potential. Compared to most other higher education systems around the world, the US system is largely independent of federal government regulations and is highly decentralized. Second, the US higher education system is considered one of the best in the world and offers flexible study opportunities at over 40,000 colleges and universities, and US degrees are recognized worldwide for their academic excellence and enhanced learning experience. And lastly it's also incredibly diverse. There are public institutions and private; very large and very small; secular and religiously affiliated; urban, suburban, and rural. Such diversity means that there is a right institution for every qualified student, and I'd like to illustrate this point by showing you the K-12 system. This is the American schooling system, and K-12 is from kindergarten to the 12th grade, which is an American expression that indicates the range of years of publicly-supported primary and secondary education found in the United States.

US national security depends on the ability to properly train and retain the next generation of Americans to safeguard their national security. This is all based on orders founded on freedom, democracy, law, and private commerce, and that's why national security strategy focuses on four actions: protecting the American people, the homeland, and the American way of life; promoting American prosperity through economic growth and fair trade, protection of intellectual property and energy dominance; preserving peace through military strength and, lastly, advancing American influence to protect the interests and principles of the US in the international arena.

Some would say that the most destructive force in America today is public opinion. Without higher education, the situation could easily be worse. Higher education increases the potential for people to perform as citizens, and US failure to educate its students threatens the country's ability to thrive in a global economy and maintain its leadership role—so finds a new council on foreign relations-sponsored independent task force. The report notes that, while the US invests more in K-12 public education, too many other developed countries' students are ill-prepared to compete with their global peers. So, though there are many successful individual schools and promising reform efforts, the national statistics on educational outcomes are disheartening. Secretary of State Condoleezza Rice said that the crisis in K-12 education is our greatest national security crisis today... education failure puts the United States future economic prosperity, global position, and physical safety at risk, and the country will not be able to keep pace globally unless it moves to fix the problems it has allowed to fester for too long.

And what I want to mention is that more than 25% of students fail to graduate from high school in four years. In Civics, only a quarter of US students are proficient or better on the national assessment of educational progress, although the US is a nation of immigrants. Roughly 8 out of 10 Americans speak only English; a decreasing number of schools are teaching foreign languages; and, lastly, more college students need to take remedial courses. This lack of preparedness poses threats on five national security fronts: firstly, economic growth and competitiveness; physical safety; intellectual property; US global awareness; and, lastly, US unity and cohesion.

Too many young people aren't employed in an increasingly-high skilled and global economy, and too many aren't qualified to join the military because they are physically unfit or have criminal records or have inadequate levels of education. The task force proposes three policy recommendations which I'm going to talk about. The first recommendation is implementing educational expectations and assessments in subjects vital to protecting national security. States should expand the common core state standards, ensuring that students are mastering the skills and knowledge necessary to safeguard the country's national security. Secondly, make structural changes to provide students with good choices. Enhanced choice and competition in an environment of equitable resource allocation will fuel the innovation necessary to transform results. Lastly, launch a national security readiness audit to hold schools and policy makers accountable for results and to raise public awareness. There should be a coordinated national effort to assess whether students are learning the skills and knowledge necessary to safeguard America's future security and prosperity, and the results should be publicized to engage the American people in addressing problems and building on success. According to our study, "also improving the education may be our greatest national security challenge ... [and] a smarter, better workforce is not a nice-to-have; it's a must-have".

The task force talked about this subject, and they are focused on three main fronts: firstly, it is rigorous standards and aligned assessments. As high standards of performance are the foundation for excellence in the military, Gregorio's academic standards are the backbone of Tennessee's K-12 public schooling. Second is high quality teaching and leadership, which I think don't need an explanation. The last main front are innovative approaches to college and career readiness. Both education and military leaders agree that innovative approaches to ensuring preparedness for college and career will be crucial for securing America's future.

So, finally, I'm going to talk about what are the links between them. Retired Admiral William McRaven, a former US navy seal who oversaw the raid on bin Laden, said that he was the biggest fan of the younger generation of Americans and that education in grade school played a broad role in national security: "it was because I recognized that unless we are giving opportunity and a quality education to the young men and women in the United States, then we won't have the right people to be able to make the right decisions about our national security." So we have got to have an education system with the US that really does teach and educate young men and women to think critically, to look outside their microcosm, because if we don't develop these great folks, then our national security in the long run may be in jeopardy.

They won't have an understanding of different cultures; they won't have an understanding of different ideas; and they won't be critical thinkers. Human capital will determine power in the current century, and the failure to produce that capital will undermine Americans' safety:

> Large, undereducated swath of the population damage the ability of the United States to physically defend itself, protect its secure information, conduct diplomacy, and grow its economy.

[See Appendix for corresponding PowerPoint presentations.]

12

An Overview of Emerging Technologies: A US Coast Guard Cadet Panel

Moderator:
- Dr. Angela G. Jackson-Summers, USCGA

Cadet Panelists:
- 1/c Abby Nitz
- 1/c Erin Wood
- 2/c Chase Jin
- 2/c Michael Dankworth
- 2/c Branyelle Carillo
- 1/c Nick Epstein

As presented at the 2022 Strategic and Security Studies Symposium
Hosted by the Institute for Leadership and Strategic Studies
University of North Georgia

Dr. Eddie Mienie:
Please join me in welcoming the moderator of our next panel—our final panel for the conference—all the way from the US Coast Guard Academy. Angela Jackson-Summers is an Assistant Professor of Information Systems in the Management Department at the US Coast Guard Academy. She received her Ph.D. in Business Administration (Information Systems) from Kennesaw State University. Her research interests include IT/IS risk management and data/information security and assurance.

Dr. Angela G. Jackson-Summers

Thank you all so much for having us join you today. We are very excited for this opportunity. The title of our presentation is "An Overview of Emerging Military Technologies: A US Coast Guard Academy Cadet Panel Presentation."

For our presenters, we have six panelists, each covering one of the emerging military technologies that were identified in a report published in November of 2021. We have for our artificial intelligence, First Class Abby Nitz; directed energy weapons, Second Class Chase Jin; lethal autonomous weapons, Second Class Branyelle Carillo; biotechnology, First Class Erin Wood; hypersonic weapons, Second Class Michael Dankworth; and quantum technology, First Class Nicholas Epstein.

Just to give some background as to what actually got us to this point, the motivation for our group research was really to broaden our awareness and learning of information technology in the military. We were fortunate enough to come across a report published by the Congressional Research Service, and that report, titled *Emerging Military Technologies: Backgrounds and Issues for Congress*, was published in November of 2021, actually November 10th, and it covered these specific six emerging military technologies.

For our class, this course is actually taught as a course in information technology in organizations, and the focus of this has been military primarily, and we wanted to be able to speak more to how the emerging military technologies could be considered or used in different missions associated with the United States Coast Guard. Each group who performed research was given these particular five missions as mentioned to consider as they actually went forth and did their work. They were also expected to describe the potential challenges or drivers that could play a role in the actual operational effectiveness and use in those missions.

To give them directed study efforts, there were two specific research questions considered for this particular work. The first research question was for them to address: what homeland security missions, of the five that I mentioned previously can the emerging military technology be used for? And then rank them in the order of most-to-least use, and actually state or substantiate why. The second research question: what challenges might exist in driving the emerging military technologies' operational effectiveness for use in achieving each homeland security mission?

Now to speak a little bit to the methodology that the class has been actually undertaking. We went through the effort of performing a literature review, and the literature review framework was based on an approach that was captured from Levy and Ellis (2006) [*A Systems Approach to Conduct an Effective Literature Review in Support of Information Systems Research*]. This was the actual framework as you see here as a depiction of it where you actually go out and identify your sources as inputs, and then you process them using Bloom's Taxonomy [Benjamin Bloom et al., 1956; *Taxonomy of educational objectives: The classification of educational goals. New York: David McKay Company, 1956.*]. One, they were expected as a group to know and comprehend the literature and then go through and apply the understanding that they captured through that reading and analyze it associated with addressing those particular research questions. Let me just say this, they are still continuously working on this effort, but you are getting a snapshot of what they have found thus far. They have gone through and actually performed some synthesis of the literature thus far and are working towards further evaluation and conclusion of efforts. So, at the end, their efforts will actually render a paper they have to submit of their compilation of efforts and perhaps also future publications throughout the next academic year.

With that said, I will start out by introducing the first panelist who is going to cover artificial intelligence, First Class Abby Nitz.

1/c Abby Nitz:

Thank you, Dr. Jackson-Summers, and what an honor it is. I think I speak for all the cadets that are going to be speaking at this panel: what an honor to be speaking at this symposium. I'll be presenting on artificial intelligence along with my group members as shown on the screen, as you can see. We have been researching this all semester.

Artificial intelligence—what really is this? Basically, we attached a definition on the screen, but, along with that, it's really just technology that's thinking like a human would, right? It's doing things that you wouldn't think technology would be able to do before. Through our research, our key points that we've found are that, essentially, artificial intelligence has led to some fast-reaction-time human operators. It can, therefore, achieve things that humans might take a much longer time to do, especially in the DoD and DHS. We're doing all these sorts of missions, and if we can have the aid of artificial intelligence, it'd be incredibly beneficial. And it can also analyze a heck of a lot larger quantity of data than any sort of human brain can do. Furthermore, the DoD has unclassified investments of $874 million for the year of 2022 in this specific area of technology. In our sector, the DHS and [within] the US Coast Guard, the overall importance that we're seeing with these new technologies and these new emerging AI fields is going to be through, especially, voice recognition and contact tracking, which we'll show in the next slide show.

These are the missions that we specifically found will be benefited from incorporating artificial intelligence into the Coast Guard. First and foremost is drug and migrant interdiction. Artificial intelligence is going to be huge when they incorporate it into drones, unmanned aerial vehicles, and unmanned surface vehicles, to track contacts. This past summer, I was actually sent down on Temporary Duty (TDY) to be in Puerto Rico doing these missions, drug and migrant, and we had aid from HC-144 planes.

An Overview of Emerging Technologies: A US Coast Guard Cadet Panel

They have these really high-def cameras that can actually find the migrants and the drugs, then end up tracking them. Once we get more, they can grasp onto an actual imagery and then trace that which is hard to lock onto when you're in the ocean, right? Looking for people that are moving—it's really difficult. So, also in law enforcement, there are non-lethal weapons that can stop vessels and vehicles without unfortunate consequences, and that's going to be some sort of huge project that we're working on and looking into. And then ports, waterways, and coastal security: there's unmanned surface vessels to conduct some harbor patrols. This is going on, a little hand-in-hand with the drug and migrant interdictions. The patrolling is going to be the huge part. We also had a week ago Rear Admiral Shofield who came and spoke to some of the cadets about some of these emerging technologies. There's a Project Maven in the works for the DoD and DHS, and that's going to be incorporated to the national security cutters and 144s, which is this computer vision that can grasp onto the imagery using AI technology.

Potential challenges that can really arise from using AI is the privacy of PII for surveillance of vessels. Regulatory and legal issues: there's really going to need to be some sort of laws with AI, which can get dicey in this specific area. And with criminal organizations obtaining their own technologies: drones, autonomous vessels, and submersibles as well as with us expanding (the DoD and the DHS getting better), other countries are going to be doing the same. Just keeping everyone in check and making sure that there are regulations and laws put in place to make sure that they're all set in stone. So, being cheap and easily available and untraceable, these technologies need to be high def, and they need to be quantifiable in that area. However, there's going to be human error, right? We can't rely on the artificial intelligence to take over for us. We have to keep our own intelligence by also using our personnel, which is the most important part of any service, especially the Coast Guard, and train a competent workforce to handle technology. That's going to be something that's going to be a battle of its own, because with us and with this new technology we don't really know. There's a lot up in the air with it. Training workers, who can use it, is going to be huge, and then another challenge is the collision and capsizing of autonomous vessels, which doesn't stop with human error. They go hand-in-hand. So, that's pretty much artificial intelligence as an emerging military technology for the Coast Guard.

Dr. Angela G. Jackson-Summers:
Thank you so much.
Alright, next we'll have biotechnology with First Class Erin Wood.

1/c Erin Wood:
Hello, everybody. My name is First Class Wood. I'm going to talk to you about biotechnology and its relevance within the Coast Guard.

To start off, you guys probably know biotechnology is basically biology, which is the study of life, and technology. With the first bullet, it's harnessing cellular and biomolecular processes to develop technologies and products to help to improve our lives and the health of our planet.

When we looked into the congressional research survey findings, they basically said biotech has a lot of great things that can be said about it and its impact. It has an impact on the military and global politics while it can be considered as threat actors creating weapons of mass destruction. There are some countries that are really trying to use biotech to create weapons, and in order for us to keep up with this, we need to be able to use it in those ways. It can also have some medicinal value. Understanding how biotech technology can possibly be used to help people heal, and in medicinal ways.

An Overview of Emerging Technologies: A US Coast Guard Cadet Panel

The way that it's relevant for the Coast Guard is that it helps with our defense readiness, law enforcement, personal performance, and living marine resources and oil spill response.

Going into it with the first research question, as this study is changing, my group actually found that instead of law enforcement we actually think it applies more to port, waterway, and coastal security, so I'm going to talk about that instead of law enforcement. But we'll start off with defense readiness.

We actually found that biosensors have a significant impact on defense readiness. One way that they can be used in the Coast Guard is that you can put biosensors on people, and they can detect certain things, like how much someone's sweating and how they're changing—physiological factors. They can determine when someone reaches a point where they're no longer at their best performance, and they need rest or things like that to just try and make the members of the Coast Guard and the military a little bit more effective and make sure that they're at their best health.

For ports, waterways, and coastal security, biosensors can be put into materials like glass or plastic. This can be helpful in border patrol when you're trying to see if illegal drugs or anything's coming in across the borders of the US Having biosensors built into some of these materials can really help with the effectiveness of tracking materials.

It's also helpful for drug interdiction because biosensors can be used in the Coast Guard. They could be put into the tips of gloves so that you can rub your hand over a surface and see if a specific drug is there, what type of drug it is, or if there's residue of it. They can also be used in forensics in this sense. It really has a lot of applicability to the Coast Guard and to probably other military services.

Potential challenges that we found that happen for biotechnology: you can see them listed here [see slide show], but the main ones I'm going to focus on is, just like most technology, they're very touchy. One thing that we found out for biotechnology is a lot of sensors actually have to be kept in controlled environments, in terms of temperature and humidity, to be the most effective.

One other thing that is really a challenge for biotechnology is that, if it needs to be in a specific environment and you're trying to use it out on a boat or during an inspection and it's in the wrong environment, it's not as effective. Then is it really worth using? That's something that's really important to know.

Obviously like all technology, cost effectiveness is a major factor. Biotech is expensive, and proper funding needs to be put into it in order for it to be successful.

Also, certain technology, specifically, biotechnology, if you're using a biosensor, sometimes it takes a long time for the data to truly come through, so one of the challenges is, we need to make biosensors that are fast reacting. If someone's on an interdiction and they rub their hand over a surface to try and see if there's any drug residue, it needs to immediately tell them, as opposed to having to wait multiple minutes or hours for some feedback. That's not real helpful when you're on the job on the mission. Thank you!

Dr. Angela G. Jackson-Summers:

Thank you so much, First Class Wood.

Now we will go into the directed energy weapons, Second Class Chase Jin.

2/c Chase Jin:

Thank you, Dr. Jackson-Summers. Our group's topic was directed energy weapons, or DEWs for short.

These are defined as weapons using concentrated electromagnetic energy, rather than kinetic energy, to incapacitate, damage, disable, or destroy enemy equipment, facilities, and/or personnel. Some examples include high-energy lasers or non-lethal high-powered microwave weapons.

An Overview of Emerging Technologies: A US Coast Guard Cadet Panel

Some key points addressed by the CRS report include concerns about technological maturity, since DEWs have questionable usability in early stages, but, despite these concerns, the report goes on to detail DEW interest and testing from the US Navy, Army, and the Air Force, as well as the DoD requesting 578 million dollars for DEW research. With the increasing focus on DEWs from the other armed services, the question becomes "What Coast Guard homeland security missions can benefit from the implementation of directed energy weapons, and what challenges may lie in the process of integrating DEWs into Coast Guard operations?"

Regarding use considerations for DEWs and the Coast Guard, let's start off with the ports, waterways, and coastal security, which deals with the defense of national critical infrastructure. The precise attacks and the minimal collateral damage of high-energy lasers as well as the ability of high-powered microwaves to disable electronics in a localized area can make these weapons the hardest hitters in port security.

Drug and migrant interdictions can also benefit from these weapons, and disabling semi-submersible narco subs. But where we truly see the potential for high-energy lasers here is the capability to wirelessly recharge vehicles by converting laser beams back to electricity. Current technology already allows for unmanned air vehicles or UAVs to operate off of wireless recharging. Along with providing UAV support, we believe that this capability could transform Coast Guard operations if implemented in helicopters, losing fuel constraints and weight.

And last, we have law enforcement. Microwave technology can be used as a non-lethal deterrent when dealing with criminal activity, mitigating much of the risk of putting Coast Guard lives on the line in more tactical missions.

And, as with any new technology, there are some challenges to implementation. First is a property of high energy lasers called absorption and scattering where water vapor in the air can impact the effectiveness of laser beams, which is compounded during bad weather conditions. And given how the Coast Guard operates over the ocean, you can see how this might be an issue.

Next is power consumption: high energy laser beams are incredibly power efficient, costing around ten dollars per shot. However, the power generating and cooling equipment requires so much power and space that, even among navy warships, only a few are able to accommodate the needs of high-energy lasers.

And last is uncertainty. Development of directed energy weapons is certainly not a cheap operation, which prompts the question, "Does the Coast Guard really need high energy lasers and microwave technology?" We have enough trouble securing funding anyhow, but through our research, we can see that directed energy weapons can indeed provide invaluable support for Coast Guard missions. Thank you.

Dr. Angela G. Jackson-Summers:
Thank you so much.
Now we have hypersonic weapons with Second Class Michael Dankworth.

2/c Michael Dankworth:
Hello, everyone. My name is Second Class Michael Dankworth, and I'm glad to be talking with you about hypersonic weapons.

Hypersonic weapons fly at speeds of at least Mach 5 and are able to change course during a flight. They are different from ballistic missiles which can also travel at hypersonic speeds of at least Mach 5 but have trajectories and limited maneuverability.

Two categories of hypersonic weapons are hypersonic cruise missiles and hypersonic glide vehicles. Hypersonic glide vehicles are launched from a rocket. It then separates from the rocket and glides at speeds of at least Mach 5 toward a target.

So, what is the significance of hypersonic weapons? It is the ability to launch weapons at hypersonic speeds giving any country a considerable advantage because such weapons can evade just about any defense system we're currently using.

How do hypersonic weapons relate to the US Coast Guard? They aid in defense readiness to protect the homeland, installed on cutters to assist with naval missions and deter attacks on ports and waterways.

In defense readiness, as previously mentioned—the Coast Guard falls under the Navy in wartime. The Navy is within the Department of Defense, and delivering hypersonic weapons is one of the Department of Defense's highest priorities. The DoD, Department of Defense, is working in collaboration with industry, government national laboratories, and academia to field hypersonic warfighting capability since the early to mid-2000s. Hypersonic missiles can help make us equal in firepower to the Navy.

The next mission it helps is ports, waterways, and coastal security. In armed conflict, we must defend allied ports, and these missiles could be used to eliminate various targets. Since we defend ports from enemies, we can use these missiles to strike against them.

Lastly, drug interdiction: regarding drugs, we could use them to threaten drug runners with the missile to help them get people and threats to stand down.

Potential challenges regarding hypersonic weapons are communication and maneuverability. Communication—basic operations like communications become significant during hypersonic flight. Personnel need continuous connectivity to operators and decision-makers through global communications and sensor systems that can operate within these high-speed environments.

And then maneuverability, hypersonic systems are designed to operate in contested environments and must be capable of overcoming a wide range of defenses. At hypersonic speeds, maneuverability is a big challenge that demands extensive calculation and development. Thank you so much for your time.

Dr. Angela G. Jackson-Summers:

Thank you so much Second Class Dankworth.

Now we have lethal autonomous weapons covered by Second Class Branyelle Carillo.

2/c Branyelle Carillo:

Hello, I'm Second Class Branyelle Carillo, and today I will be presenting our group's research topic, which is Lethal Autonomous Weapons.

What is a lethal autonomous weapon? It's a special weapon that is able to operate and engage a target without human intervention.

There are some debatable key points that we found in the CRS report. The first is the definition of LAWS. Countries like China, Russia, as well as the US, have had different definitions for what autonomous weapons are; therefore, there will need to be an international agreement on the definition of LAWS.

The next debate was whether or not we should allow semi-autonomous weapons instead of autonomous weapons, because semi-autonomous weapons allow humans to be involved with the operation versus autonomous weapons that are solely just the weapons themselves.

Then, lastly, the debate of who would be ethically responsible for weapons. The military in general has a hierarchy or, rather, a chain of command to place responsibility or have that responsibility, so the ethical responsibility of who is responsible if something were to go wrong with the autonomous weapons is unclear. That's why 30 countries and about 165 non-governmental organizations are calling for a ban.

Why is this important to the Coast Guard? It's very applicable to the Coast Guard, because it can be incorporated into the coastal security and deployable forces.

With that being said, we think that, based on our missions, LAWS could be very applicable to us because they are very advanced, and they are able to just solely do the mission if we needed it to be done versus having all those human factors. So, the only thing is, though, we think that it will be difficult to actually implement LAWS within the Coast Guard because we are a humanitarian service. We think that it would be better for us to have a non-lethal type of weapon system as part of the Coast Guard to fulfill these roles within our missions.

The potential challenges that we saw were, again like I mentioned earlier "Who is going to be ethically responsible for these LAWS if something were to go wrong?"

Training was another one, because these weapons are coded and programmed to solely do one operation, one challenge, or whatever the case may be.

And then the cost, it would be very expensive for us to provide these lethal autonomous weapons for the Coast Guard. Thank you for listening.

Dr. Angela G. Jackson-Summers:

Okay, thank you so much, Second Class Carillo.

Now we have our last emerging military technology, quantum technology, from First Class Nicholas Epstein.

1/c Nicholas Epstein:

Thank you, Doctor. I'm going to get right into it.

Quantum technology, so, we have a definition here [in slide show] that speaks to specifically how quantum technology is referenced, but at its root what quantum technology is, is applying quantum theory and quantum information science to technology and computers as we use them today. What that means, without getting totally into it, is the application of qubits. I'll talk a little bit more about that as we move further into what quantum technology can be applied to in the Coast Guard and the challenges with it. Specifically, information security, intrusion detection and protection, those are three major things that quantum technology can have an advantage with.

Moving forward, specifically, these three missions for use considerations are what we regarded as the most important throughout our research. Ports, waterways, and coastal security, defense readiness, and law enforcement. Each one of these is regarded, similarly in the way, that quantum technology would be used, and that is due to the application of quantum computers in building both intrusion detection and intrusion prevention systems. Each one of these mission avenues relies on data security and communication protocols. Having the more advanced version of an intrusion detection system using machine learning with the capabilities of quantum computing would be huge for defending each one of these missions to prevent similar attacks to the warcry tactics.

Potential challenges, the biggest one for quantum computing right now is getting not only funding, but also backing and people trusting that this quantum technology is coming soon. So being able to produce it in a timely manner, not making these parties that are investing in it wait, not just DoD but the private sector as well. Reducing the pushback in the use of it (in terms of its ethics and how it should continue to be used), makes one think if it is ethical at all.

Quantum technology, as a whole, presents a massive opportunity to increase the speed of crypto analysis simulations, preventing terrorist attacks, and predicting the likelihood of outcomes in almost all scenarios with the use of qubits and all quantum technology.

Thank you for your time, and I'll hand it back over to Dr. Jackson-Summers.

Dr. Angela G. Jackson-Summers:

Thank you so much.

I'm very proud of all of the efforts that these groups have performed to date and realize that they have really gone over and beyond from the standpoint of giving a different perspective just to a number of these technologies.

So, what you have and what you heard:

- Each one of them defined their understanding of these emerging military technologies.
- They captured key points that were addressed within the actual report itself.
- And they gave some description of the importance of these emerging military technologies to the Coast Guard, as well as really going through and ensuring that they responded to the research questions that were given.

We believe overall that the contributions of these particular group efforts will create future discussions in the advancement of emerging military technology, especially from the standpoint of gaining further knowledge within the US Coast Guard environment as it relates also to its missions. It will help us to advance our learning opportunities, basically, when we think about broadening our awareness of how information technology in the military is used. Lastly, they have fostered collaborative leadership growth that they are currently performing to date by doing problem-solving and increasing or advancing their communications and critical thinking skills.

With that, that concludes our panel presentation.

Appendix

How Higher Education Fills the Security Gap in the Post-Cold War Era

Dr. Craig Greathouse

Graduate Education and National Security: Expanding Understanding

Craig B. Greathouse Ph.D.
Professor of Political Science
University of North Georgia
U.S Higher Education and National Security Conference
April 6-7, 2022

Agenda

- Graduate education defined
- Importance of M.A. level
- Disciplinary Needs
- Broader Impact/Bigger Picture
- Problems
- Conclusion

Graduate Education Defined

- Education beyond the level of the bachelor leading to a recognized and accredited Masters or Doctorate level degree

- Certificates and other forms of post baccalaureate education provide broadening of skills but may not translate to expanded understanding of data and/or critical thinking capacity

- Problems is online scam schools

Importance of M.A. level

- For most people the doctorate degree is not needed and can actually undermine career progression

- Doctorate programs focus on creation of knowledge and understanding the cutting edge

- Cost, skill, and time commitments to doctoral level education make it not viable for the numbers that need additional education

Importance of M.A. level

- M.A. builds on basic skills acquired at the bachelors level

- Expands critical thinking, writing, and analytical skills

- Provides understanding of methodology, use of data, and integrated research projects

- Introduction to the use of theoretical frameworks and other heuristic devices to allow for consistent analysis

Disciplinary Needs for National Security Analysis and Understanding

- International Relations / Affairs
- Strategic and Security Studies
- Political Science / specifically comparative politics & public policy
- Geography/GIS
- Sociology / Anthropology
- Economics
- Computer Science / cyber security
- Modern History
- Language

Disciplinary Needs For National Security

- Certain MA Degrees lend themselves to seeing the broader picture within the system
    - International Relations / Affairs
    - Economics
- Certain MA degrees are necessary for understanding specifics
    - Strategic Studies
    - Comparative Politics / Public Policy (international)
    - Sociology / Anthropology
    - Computer Science / cyber security
    - Modern History
    - Language
    - Geography / GIS

Bigger Picture / Broader Impact

- A big picture in understanding systemic influences along with specifics with state and non-state actors is critical

- Understanding the bigger picture allows for analysis that can be supplemented and expanded by specific information but without big picture understanding mistakes happen
    - Argument by many that NATO should initiate no-fly zone over the Ukraine during the 2022 conflict

Example

- M.A. in International Relations/Affairs
    - Specific example M.A.I.A. at University if North Georgia
    - Foundational Courses: Understanding of major actors and influences
        - US Foreign Policy
        - I.R. Theory & Theories of Comparative Politics
        - I.P.E
        - International Security
        - Global Governance
        - Research Methods
        - Capstone
    - Electives: Specifics of regions or content areas (specifically security)

Problems of Graduate Education for National Security

- Pushing flavor of the moment degrees and programs of study
    - Terrorism studies post 9/11
    - Intelligence degrees
    - Security studies
- Asserting that a degree/area of study provides a broader understanding than it does
    - Language
- Online programs that do not hold the standards
    - Canned content and or limited content
    - Limited time – all 8 week classes or life experience
    - Focused on particular groups – VA benefits and current military
    - Predatory advertising and significant cost
    - Qualification of many "faculty"

Conclusion

- Need for graduate education not certificates / badging

- Focus on M.A. level

- Combination of big picture and specific programs

- Existence of limited degrees and bad actors

HOW HIGHER EDUCATION FILLS THE SECURITY GAP IN THE POST-COLD WAR ERA
Dr. Cristian Harris

International Education & Public Diplomacy

Dr. Cristian Harris
Department of Political Science & International Affairs
University of North Georgia

Public Diplomacy (PD)

- PD is a government's effort to communicate directly with foreign publics in order to bring an understanding of its country's ideas, values, institutions, culture, and its policies.
- PD is not a uniquely state-run activity. Non-state actors can engage in PD by presenting a variety of private points of view in addition to the official government position expressed by diplomats.
- The most visible manifestations of PD are:
  - Cultural and academic exchanges, student exchanges, military exchange programs, foreign broadcasting services, cultural centers and libraries, language institutes, art exhibitions, and cultural performances.

Private Actors and Public Diplomacy

- Universities have played a formative role in U.S. public diplomacy.
- One of the first U.S. agencies engaged in PD was the Division of Cultural Studies of the Department of State (1938).
  - The DCS assisted universities in their educational opportunities overseas.
  - Scholars, scientists, and artists were encouraged to develop long term contacts with their peers overseas.
- Later efforts such as the Fulbright Program (1946) accentuated the role of universities and private actors in U.S. public diplomacy.
  - During the Cold War and particularly after 9/11, the U.S. Congress funded student exchange programs through the State Department.
  - The Rockefeller Foundation and Carnegie Corporation
  - Institute of International Education (IIE)

International Education as National Security Priority

- U.S. Information and Educational Exchange Act of 1948
- National Defense Education Act of 1958
- Mutual Educational and Cultural Exchange Act of 1961 (Fulbright-Hays Act)
- Higher Education Act of 1965
- National Security Education Act of 1991
    - Boren Scholarships and Boren Fellowships, Language Flagship, Project Global Officer (GO)

Shifting Priorities

- Universities have moved beyond the exchange of students and scholars toward Comprehensive Internationalization (CI).
    - CI is defined as a commitment "to infuse international and comparative perspectives throughout the teaching, research and service missions of higher education" (Hudzik 2011).
- Globalization pressures
    - Education is one of the most important U.S. exports of services.
    - U.S. universities face a fiercely competitive global market.
- Strategic drift and lack of consensus
    - Post-Cold War; Post-9/11

Suggested Readings

- Goodman, A.E. (2013, Fall). What is the next big thing in international education? *IIE Networker*, 7.
- Hudzik, J. K. (2011). *Comprehensive internationalization: From concept to action.* Washington, DC: NAFSA.
- King, C. (2015, July-August) The decline of international studies: Why flying blind is dangerous. *Foreign Affairs*, 94(4), 88-98.
- Melissen, J. (Ed.). (2005). *The new public diplomacy: Soft power in international relations.* Palgrave Macmillan.
- Nye, J. (2004). *Soft power: The means to success in world politics.* Cambridge, MA: Public Affairs.

HOW HIGHER EDUCATION FILLS THE SECURITY GAP IN THE POST-COLD WAR ERA
Dr. Dlynn Armstrong-Williams

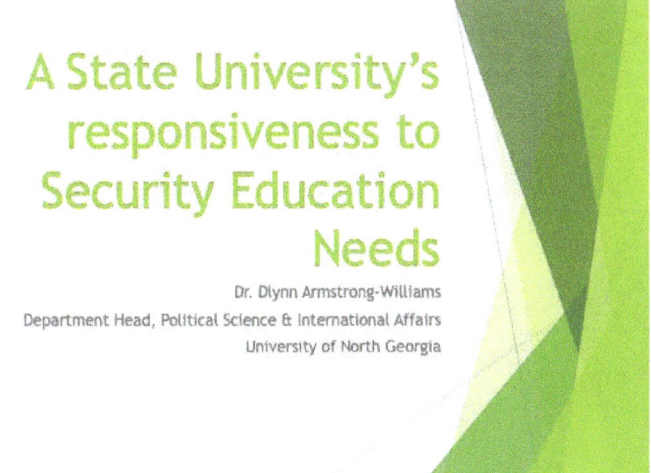

A State University's responsiveness to Security Education Needs

Dr. Dlynn Armstrong-Williams
Department Head, Political Science & International Affairs
University of North Georgia

Efforts at the then NGCSU for Curricular Modification
(Internal Institutional Context)

- 2007
- Campus population of approximately 6000 students
- Corps of Cadets approximately 500 students
- Coordination of International Efforts at NGCSU occurs within the Executive Affairs area through the CGE
- Campus wide internationalization plan launched
- Undergraduate International Affairs Degree Established

The Greater Context—HASC Committee Report

- *Building Language Skills and Cultural Competencies in the Military: DOD's Challenges in Today's Educational Environment (November 2008)*
- "There is no doubt that foreign language skills and cultural expertise are critical capabilities needed by today's military to face the challenges of our present security environment. But, only a small part of today's military is proficient in a foreign language and until recently there has been no comprehensive, systematic approach to develop cultural expertise."
- "Today's military establishment, its active duty, reserve, and civilian personnel, must be trained and ready to engage the world with an appreciation of diverse cultures and to communicate directly with local populations. These skills save lives." (Executive Summary).

Desired Outcomes Outlined in HASC Report

- The Department has personnel with language skills capable of responding as needed for peacetime and wartime operations with the correct levels of proficiency.
- The total force understands and values the tactical, operational, and strategic asset inherent in regional expertise and language.
- Regional area education is incorporated into Professional Military Education and Development. (HASC, 2008, 29).

Context-US Army Cadet Command's Cultural Understanding and Language Proficiency (CULP) Program

- Goal: To create Commissioning Officers who possess the right blend of Language and Cultural Skills required in support of global operations in the state of persistent warfare expected in the 21st century.
- 50% of all SROTC cadets to experience OCONUS culture and language immersion.

Curricular Response to Outlined Security Demands

- Construction of the Undergraduate International Affairs Degree at then NGCSU—launched in 2007
- What makes it unique:
  - General courses for all students: Comparative Security Issues, International Relations Theory, Comparative Government, Politics of Development, International Law/International Organizations, International Political Economy.
  - Students must choose a regional concentration (Asia, ME& North Africa, Europe or Latin America)
  - Students are required to complete language study up to the 2xxx level, but they are encouraged to continue language at the 3xxx level and it will fall into their regional concentration of study.

International Affairs Curriculum continued...

- In the regional area—students can take courses in POLS, HIST, INTL and LANG.
- All students are required to complete both a study abroad experience and an international internship. These experiences must coordinate with the student's chosen regional area of study and their strategic language.
- Study abroad experiences must be at least 25 days in country.
- International internships must be at least 320 contact hours with at least 60% of the student's work hours including cross-cultural communication and engagement.
  - Students are permitted to work with international partners in the United States if travel is too costly as long as the internship meets the cross cultural content.

Degree integrates HIPs and initiatives from the DOD

- Incorporates High Impact Practices (HIPs) into the degree:
  - Study Abroad
  - Internships
  - Senior Capstone Experience.
- Research shows that students involved in high-impact practices (HIPs) enjoy higher levels of learning success.

Challenges in implementation

- Fewer camp dates for cadets making it more difficult for students to participate in study abroad
- Financial challenges linked to increased costs for experiences beyond the classrooms (HIPs).
- Curricular coordination between all departments regarding course offerings and the MILS department.
- Covid-19 impacts on student and faculty mobility
- Effective communication between desired skills in PME and academic units
- Flexibility to allow for access to national scholarships, double-majors and commissioning dates.

Where we are now...

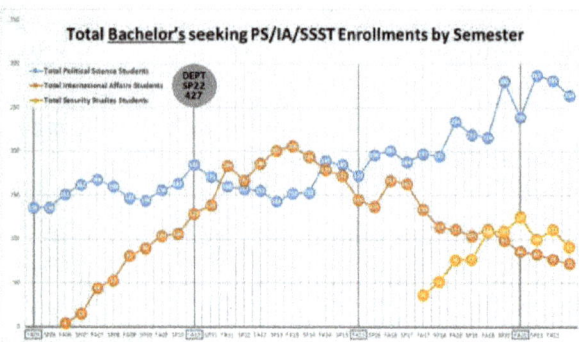

Where we are going...

▶ Security Needs are constantly shifting...presently there is ongoing discussion about the evolving needs of public diplomacy and the need for increased education regarding the Operational Informational Environment (OIE).

▶ In 2020 the US Advisory Commission on Public Diplomacy submitted a report to Congress and the Sec. of State to address the increasing complex and competitive global information environment entitled *TEACHING PUBLIC DIPLOMACY AND THE INFORMATION INSTRUMENTS OF POWER IN A COMPLEX MEDIA ENVIRONMENT.*

▶ Teaching about digital influence efforts is one aspect of a larger effort to educate military and civilian government professionals on the information environment, described in joint doctrine as "the aggregate of individuals, organizations, and systems that collect, process, disseminate, or act on information." Developing a curriculum for the information environment is an immense instructional effort; the learning requirements for comprehension of key elements of the IE as well as *operations* in the information environment (OIE) include a disparate range of practices and tools, from electronic warfare, to public diplomacy and strategic communications, to the growing array of instruments involved in cyber operations. (USACPD, 2020, 38).

Appendix

Curricular Integration to assist students in preparing for new operational environment

- Transitioning POLS 4295: Special Topics: Political Polarity course to an ongoing course within the curriculum.
- Increased concern for influence operations linked to Cyber.
- As stated in the report: *From a teaching perspective, traditional political communication concepts could be paired with emergent observations of influence attempts to trace the potential impacts of influence operations.*
- The course will continue to focus on the intensification of polarity in politics linked to the ongoing use of disinformation and will examine how influence might manifest or be conditioned in a citizenry. Students will be introduced to case studies which examine the use of disinformation, strategic communications and as well as skills needed to assess influence operations.
- Goal: New course to be included in the PME requirements at UNG.

SYMPOSIUM TRANSCRIPT
COLONEL WORTZEL
Colonel Larry M. Wortzel

Larry M Wortzel, PhD

Colonel, U.S. Army (Retired)
Senior Fellow in Asian security
American Foreign Policy Council

https://www.afpc.org/

China's Geostrategic Position

Zheng He's Voyages

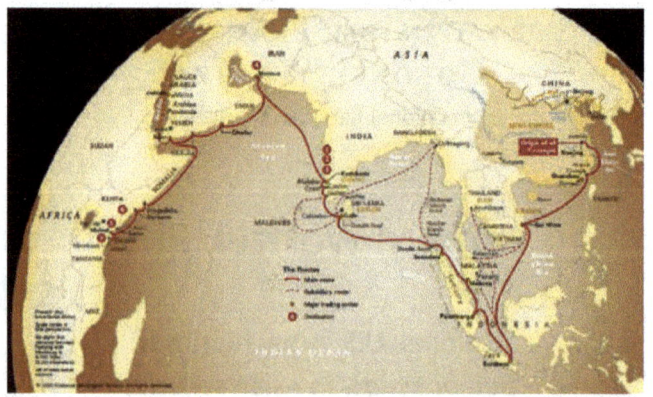

Appendix

OBOR's Maritime Route

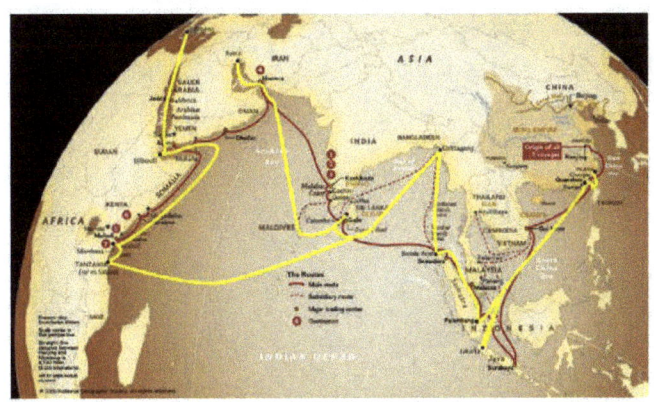

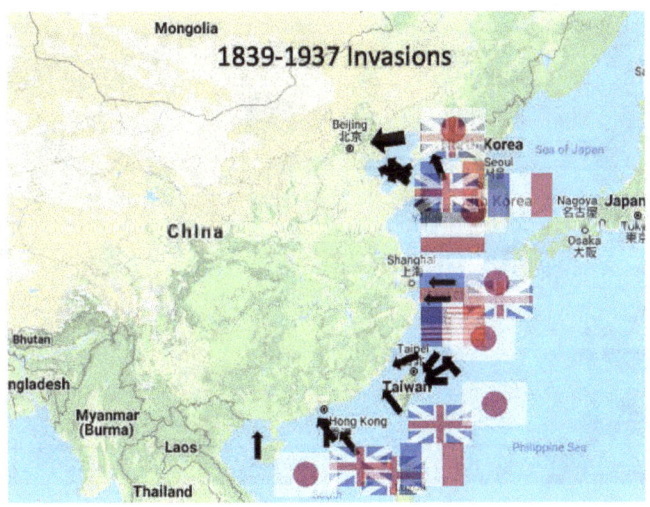

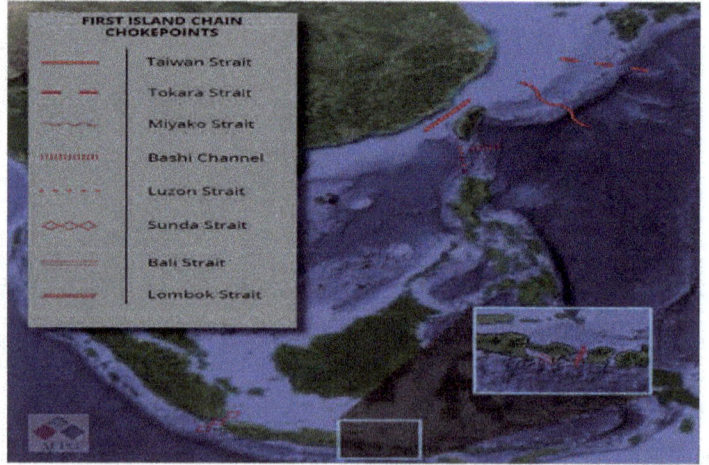

Appendix

Technologies Involved in the Competing Strategies in the Pacific

- Hypersonic missiles, warheads and maneuvering warheads
- Over the Horizon Radar
- Networked targeting systems
- Multispectral forms of target detection
- Space-based assets
- Networked, air, sea and ground based systems
- Long-range, precision strike cruise missiles, artillery, warheads
- Networked massed fires combined with electronic warfare, electronic countermeasures, and cyber attack
- Swarmed unmanned vehicles (drones)
- Artificial intelligence
- Directed Energy

RETHINKING HIGHER EDUCATION PRACTICES TO STIMULATE INNOVATION AND GLOBAL SECURITY

Crystal Shelnutt, Ed.D.

Innovation and Higher Education in the 21st Century

Innovation and Higher Education

1. Necessity of Innovation in Higher Education
2. Attributes of Innovation
3. Barriers to Implementation of Innovation
4. Conclusion

 UNIVERSITY OF WEST GEORGIA

Innovation in Higher Education

1. Collaborative skills
2. Literacies
3. Analysis
4. Application

 UNIVERSITY OF WEST GEORGIA

Competencies

1. **Intercultural awareness**
   - Pattern of human behavior
   - Function effectively
2. **Critical engagement**
   - Value systems
   - Ethical decision making
3. **Effective communication**
   - Appropriate & accurate articulation
   - Staging of self/ideas

Barriers to Innovation

1. **Organizational change**
   - Faculty governance
   - transparency

2. **Restricted view: training versus educating**
   - Mission drift
   - Nature of education

3. **Efficiencies versus transformation**
   - Scaling
   - Value propositions

Conclusions

- **Programmatic Development**
  (Kezar, 2001)

- **Partnerships**
  (Spracher, 2010)

- **Complementarianism**
  (Merisotis & Stavridis, 2022)

Thank you!

Appendix

References

Christensen, C. M. & Eyring, H. J. (2011). *The innovative university: Changing the DNA of higher education from the inside out.* Jossey-Bass.

Cross, T., Bazron, B., Dennis, K., & Isaacs, M. (1989). *Towards A Culturally Competent System of Care, Volume I.* Washington, DC: Georgetown

Fain, P. (2015). *College completion rates decline more rapidly.* Inside Higher Ed. Retrieved from https://www.insidehighered.com/quicktakes/2015/11/17/college-completion-rates-decline-more-rapidly

Hora, M. T., Benbow, R. J., & Oleson, A. K. (2016). *Beyond the skills gap: Preparing college students for life and work.* Harvard Education Press.

Kezar, A. (2001). Understanding and facilitating organizational change in the 21st century: Recent research and conceptualizations. *ASHE-ERIC Higher Education Report, 28*(4), Jossey-Bass. https://files.eric.ed.gov/fulltext/ED457711.pdf.

Kezar, A. & Eckel, P. D. (2002). The effect of institutional culture on change strategies in higher education: Universal principles or culturally responsive concepts? *Journal of Higher Education,* 73(4), 435-460.

Lumina Foundation. (2021). *Integrating work and learning in the new talent economy.*

Spracher, W. C. (2010). National security intelligence professional education: A map of US civilian university programs and competencies [ProQuest Information & Learning]. In *Dissertation Abstracts International Section A: Humanities and Social Sciences* (Vol. 70, Issue 7–A, p. 2416).

RETHINKING HIGHER EDUCATION PRACTICES TO STIMULATE INNOVATION AND GLOBAL SECURITY

Iyonka Strawn-Valcy

US DOS & DOE Joint Statement of Principles in Support of International Education

"The robust exchange of students, researchers, scholars, and educators, along with broader international education efforts between the United States and other countries, strengthens relationships between current and future leaders. These relationships are necessary to address shared challenges, enhance American prosperity, and contribute to global peace and security..."

Source: U.S. Secretary of State Antony Blinken announces Joint Statement of Principles in Support of International Education at the 2021 EducationUSA Forum.

U.S. Approach to HEI International Initiatives

- HEI Comprehensive Internationalization Strategies
    - Study in the United States by international students, researchers, and scholars
    - Study abroad for Americans
    - International research collaboration
    - The internationalization of U.S. campuses and classrooms
    - International Branch Campuses

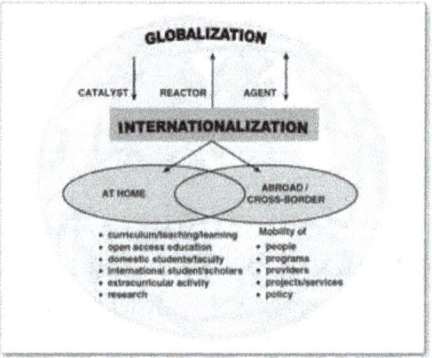

Appendix

Benefits of HEI Internationalization

- Equips participants with global and cultural competencies
- Enhances scientific discovery/innovation, and the global economy.
- Enhances cultural and linguistic diversity
- Helps in developing cross-cultural communication skills, foreign language competencies, and enhanced self-awareness and understanding of diverse perspectives.
- Strengthens relationships between current and future leaders to provide national and global leadership.
- Address shared challenges, enhance American prosperity, and contribute to global peace and security.
- Impacts how the United States is perceived globally

Threats to HEI Internationalization

- Geopolitical Events
- Economic Challenges
- Cybersecurity Threats
- Migration Crises
- Climate Change
- The COVID-19 Pandemic

"It is vital to reinforce our people-to-people relationships around the globe and to strengthen the infrastructure and pathways that help prepare Americans in all sectors to engage with the world."

HEI Strategies to Challenges and Threats of Internationalization

- Innovation in personnel and hiring: Safety & Security Officers and offices, additional compliance staff, etc.
- Emergency program funding
- Committees and work groups to address operations, risk management, and international crises
- Incentives to global learning
  - Global Learning Scholarships
- Virtual learning

Innovation in Research & Teaching

Growing importance of university partnerships, networks, and collaborative initiatives with international organizations	The significance of research to manage and address crises has become evident to policy makers and the public
• Harness international ties to support students • The development of (taking increased advantage of) branch campuses and expansion of agreements/agreement dynamics • Working with local/regional faculty to a greater degree • Virtual advising • Online tutoring	• New vaccines • Support society with related crucial projects • Top research institutions, in particular those specialized in the life sciences, may receive greater emphasis and funding • Address the United Nations' Sustainable Development Goals

Source: Jones, et al, (2021)

Appendix

HEI Strategies to Challenges & Threats to International Research Collaborations

Recognize potential conflicts of interest (COI) and conflicts of commitment (COC)

Insist on transparency, reciprocity and adherence to research integrity when engaging in collaborations or hosting a scholar

Comply with rigorous disclosure of affiliations and commitments

The 6 core values of research integrity: **Objectivity, honesty, openness, accountability, fairness, and stewardship**

Supporting HEI Internationalization & National Security

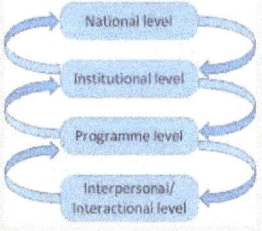

Employ a coordinated national approach to international education.	Partnerships between federal/state/local government, U.S. HEIs, schools, non-governmental entities, the business community, and others.
	Safely and securely welcome international students, researchers, scholars, and educators to the US and encourage a diversity of participants, disciplines, and HEIs.
	Encourage U.S. students, researchers, scholars, and educators who reflect the diversity of the U.S. population to pursue international experiences.
	Leverage existing international education programs and resources and create new opportunities to broaden access.

RETHINKING HIGHER EDUCATION PRACTICES TO STIMULATE INNOVATION AND GLOBAL SECURITY

Magdalena Bogacz, Ed.D.

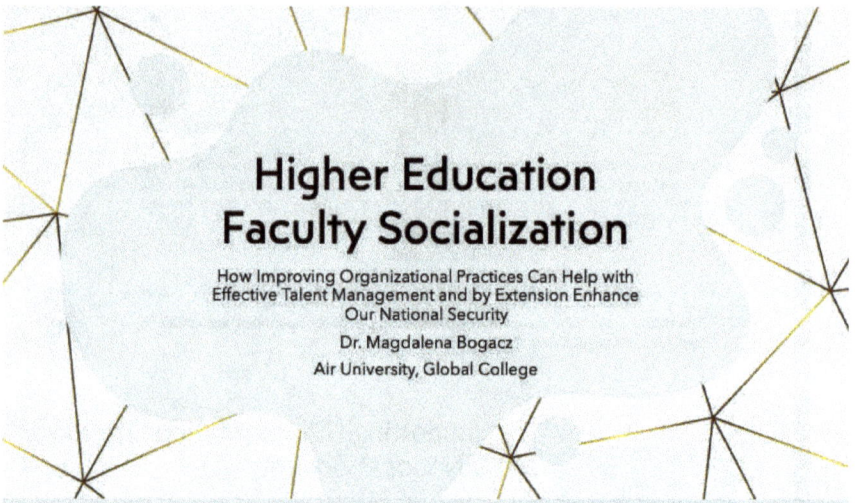

What is Faculty Socialization?

Faculty socialization refers to the process of "**how faculty learns to be faculty**" (Tierney & Rhoads, 1993).

More broadly, faculty socialization consists of mechanisms leading to the faculty transition from being organization outsiders to insiders (Bauer et al. 2007).

The process can be divided into two stages:

1. **Anticipatory stage** - occurs prior to employment, usually during graduate school
2. **Organizational stage** - occurs after employment is completed, during onboarding and new faculty mentoring programs

Problem of Practice

Current research on the topic suggests that **faculty socialization processes face challenges that disproportionately affect underrepresented populations** such as women, people of color, people differently able, and international students (Bridget, McCann, Porter, 2018). Some academic disciplines experience more faculty socializations problems than others:

1. For instance, STEM fields, but also philosophy, theology, and logic, are generally more prone to disadvantage historically marginalized populations.
2. Within these subjects, minority groups are underfunded, underrated, and under-rewarded
3. Moreover, there are disparities in the number of support groups, post-graduation placement mechanisms, as well as during onboarding and new faculty mentoring programs that take care of non-mainstream students and junior faculty.

Significance of the Problem with Faculty Socialization Practices

1. Social (In)Justice
2. (Un)Fairness of Organizational Practices and Policies
3. (In)Effective Talent Management
4. Quality of Higher Education
5. Enhancement of National Security

Faculty Socialization & Enhancement of National Security

There is a wealth of research on innovation that shows:
1. Innovation requires an organization that is open to new ideas
2. Innovation requires multidisciplinary perspectives and local knowledge of societal and human needs
3. Innovation requires diverse backgrounds
4. Higher education is currently still too dominated by silos and a lack of critical voices; therefore, its potential for innovation is currently lacking

Justification

In 2020, in the document entitled "Developing Today's Joint Officers for Tomorrow's Ways of War" the Joint Chiefs of Staff presented their new vision and guidance for Professional Military Education (PME).

"Our vision is for a fully aligned PME and talent management system that identifies, develops, and utilizes strategically minded, critically thinking, and creative join warfighters skilled in the art of war and the practical and ethical application of lethal military power"

Re-imagined PME and HE programs should rely more on **innovation, creativity, original thought, and cutting-edge research** to keep up with globalization, the return of great power competition, and the constantly changing character of war.

Solutions – Actionable Deliverables

To achieve intellectual overmatch against adversaries we must improve our faculty socialization practices:
1. Equitable training and mentoring of students and junior faculty (self-efficacy)
2. Shifting mindsets and relocating organizational resources to better help those that need it the most → valuing and fostering cognitive diversity (attainment value)
3. Creating an academic culture that cares about justice and fairness of organizational practices

Is there anything the military can do to support higher education?

1. Create, support, and fund initiatives that encourage collaboration between private and public sector that foster innovation and advance critical voices

Implications

Improved faculty socialization practices would increase cognitive diversity among higher education and professional military education faculty and by this provide a more comprehensive learning experiences and thus, generate a more comprehensive knowledge about security environment.

By including more diverse voices among our educators we would gather, assess, analyze, evaluate, and disseminate information in a more **inclusive, global, and complete fashion** that would better align with the new vision of education as presented by the Joint Chiefs.

SYMPOSIUM TRANSCRIPT
DR. MARGARET E. KOSAL

Dr. Margaret E. Kosal

National Security, Emerging Tech & Higher Education

Margaret E. Kosal, PhD
Associate Professor
Sam Nunn School of International Affairs
Georgia Institute of Technology
margaret.kosal@inta.gatech.edu

Appendix

From Science to Security

- Ph.D., Chemistry UIUC
  - Bio-inorganic chemistry and nano-structured materials
- Co-founder ChemSensing Inc.
  - Explosives, biological & chemical agent detection
- Fellow at Stanford University's Center for International Security and Cooperation (CISAC)
- S&T Advisor in Office of the Secretary of Defense
- Senior Advisor to Chief of Staff of the US Army
- Vice Chair/Study Member, US National Academies (NAS) Committees on Assessing and Improving Strategies for Preventing, Countering, and Responding to WMD Terrorism: Chemical Threats/Biological Threats

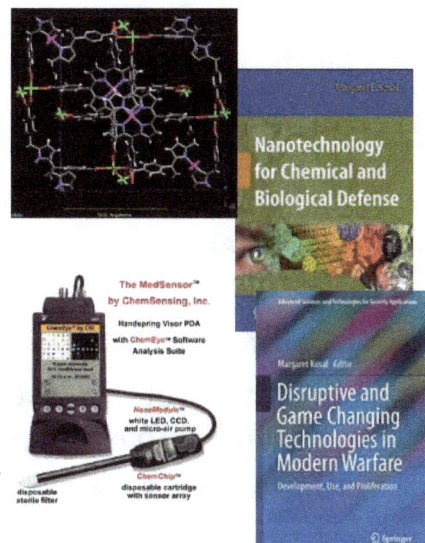

Policy & Technical Advisor, US DoD

- S&T Advisor to the ATSD(NCB) - OSD
  Interface with other DoD offices and interagency groups
  - Nonproliferation and Arms Control Technology Working Group (NPAC TWG)
  - National Nanotechnology Initiative (NNI)
  - Counterproliferation Review Committee (CPRC)
  - National Science Advisory Board on Biosecurity
  - NATO RTO
  - Liaison to DTRA-CB

  Chairing the Nanotechnology for Chem-Bio Defense 2030 Study & Workshop
  Coordinating CBDP basic research efforts, including the basic research strategy

- Advisor to Chief of Staff of the Army
  Design, organization, & employment of future Army
  Joint & conventional-SOF interoperability
  Modernization priorities
  Counter-WMD force structure & capabilities

Outline

- Strategic & Security Contexts
  - Security Puzzles
  - Motivation
- Geopolitics
  - ML & AI
- Adoption
  - Neurotech - BCIs
- Threats & Countermeasures
  - Adaptive Camouflage (nanotechnology)
- Conclusions & Recommendations

Technology's Role in Politics & War
... & the Role of Politics & War in Technology

Peter Turchin, et al., **War, Space, and the Evolution of Old World Complex Societies**, *Proceedings of the National Academy of Sciences (PNAS)*, 2013

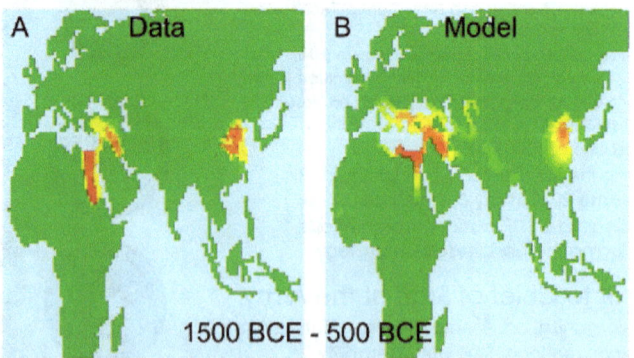

"Comparison between data and model predictions for three historical eras [one shown]. Red indicates regions that were more frequently inhabited by large-scale polities, yellow shows where large polities were less common, and green indicates the absence of large polities."

Appendix

"Less predictable is the possibility that research breakthroughs will transform the technological battlefield. Allies and partners should be alert for potentially disruptive developments in such dynamic areas as information and communications technology, cognitive and biological sciences, robotics, and nanotechnology.

The most destructive periods of history tend to be those when the means of aggression have gained the upper hand in the art of waging war."

http://www.nato.int/strategic-concept/index.html May 2010

Security Puzzles

- How do emerging technologies affect international security and the security dilemma? And vice-versa?
- Do emerging technologies – *biotechnology, robotics, AI, additive manufacturing, nanotechnology (including meta-materials), cognitive neurosciences, quantum, hypersonics* – have unique strategic value?
- How do these technologies affect conflict and cooperation?
- What are the identifiable technical (material & knowledge), structural (organizational), & political (ideational) factors? How do they interact?
- Perception or ideations
  - "Hope" & "Horror" hype
  - Rhetoric
  - (Pseudo)-technical assessments
  - Influencing factors: institutional, ideational, organizational
- Best approaches to governance – domestically & internationally – of dual use technologies and contentious research
  - Adequacy of traditional arms control treaties
  - Value of norms
  - Role of NGOs, transnational actors, industry
  - "Good neighbor" risks

AI & Geopolitics

- Statements & rhetoric
  - Hype
  - Warnings

- Who has advantage in AI battle?

What is AI?

2019 National Defense Authorization Act
- Any artificial system that performs tasks under varying and unpredictable circumstances without significant human oversight, or that can learn from experience and improve performance when exposed to data sets
- An artificial system developed in computer software, physical hardware, or other context that solves tasks requiring human-like perception, cognition, planning, learning, communication, or physical action
- An artificial system designed to think or act like a human, including cognitive architectures and neural networks
- A set of techniques, including machine learning, that is designed to approximate a cognitive task
- An artificial system designed to act rationally, including an intelligent software agent or embodied robot that achieves goals using perception, planning, reasoning, learning, communicating, decision making, and acting (NDAA 2019, 1697-1698)

Appendix

Military Applications of AI

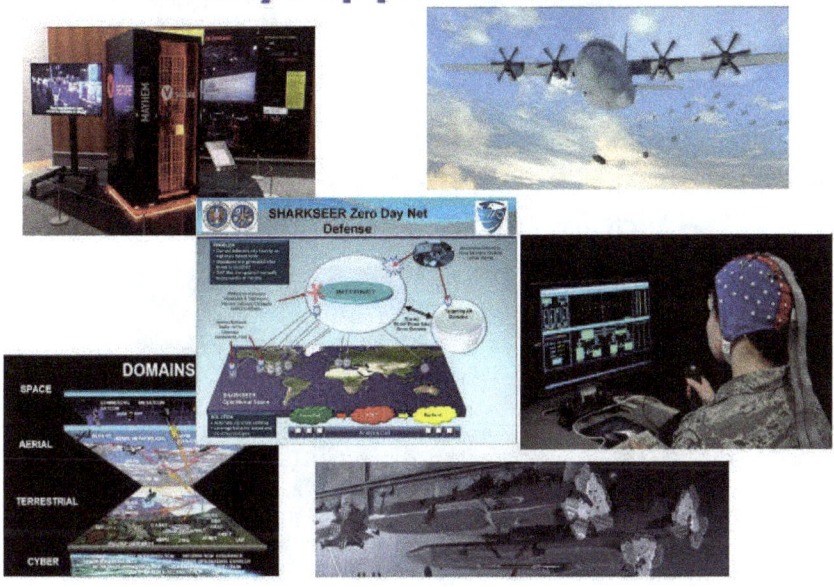

General vs Narrow AI

General AI
- Describes a computer intelligence that surpasses human intelligence across the breadth of human capability
- Includes complex decision making and critical thinking
- Several decades in the future

Narrow AI
- Programs designed to solve a specific problem: navigation, image recognition, language translation, etc.
- Includes multiple different techniques: machine learning, neural nets, big data
  - These terms are often are used interchangeably with AI, but are not AI
  - These things describe the "how" of Narrow AI. (US AI committee 2019)

> "Artificial intelligence is the future, not only for Russia, but for all humankind. It comes with colossal opportunities, but also threats that are difficult to predict. Whoever becomes the leader in this sphere will become the ruler of the world."
> – Vladimir Putin, September 2017

https://www.rt.com/news/401731-ai-rule-world-putin/

Hype ?

CATGENIE   SUPPLIES   ACCESSORIES   WHY A.I   ABOUT US   CONTACT US   BUY NOW

What is CatGenie A.I.?

The CatGenie A.I. is an IoT (Internet of Things) appliance that uses artificial intelligence (A.I.) technology to monitor your cats' bathroom behavior, keep an eye how your CatGenie is running and let you know when supplies are running low.
This data is analyzed by the CatGenie A.I. and displayed in the CatGenie App. It uses "machine learning" to help your unit clean more efficiently, monitor your cat's health and wellness, as well as create a cleaning schedule customized to suit you and your cat.

BUY NOW

Appendix

Russia's 'Status-6'

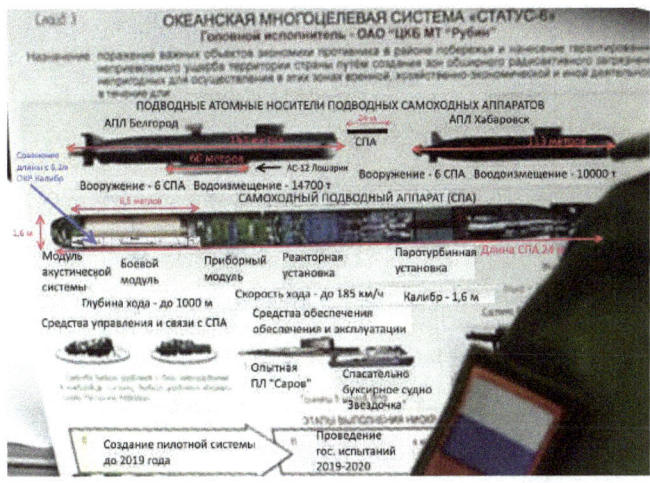

Image Credit: "Madnessgenius"

Discourse

Elite Statements

- [Nanotech is ...] "the key to developing new, modern and effective military systems"
- "Russia's economic potential has been restored, the possibilities for major scientific research are opening up. The concentration of our resources should stimulate the development of new technologies in our country. This will be key also from the point of view of the **creation the newest, modern and super-effective weapons systems.**"
- "Nanotechnologies ... [can] radically change our concepts of modern warfare."
- "A new round of the arms race is developing in the world."

"FOAB" thermobaric bomb enabled by nanotechnology

'Glory to Nanotechnology'

PRC AI Strategy

- "New Generation Artificial Intelligence Development Plan (AIDP)"
- Goal to be global AI lead by 2030
- "AI has become a new focus of international competition. AI is a strategic technology that will lead in the future; the world's major developed countries are taking the development of AI as a major strategy to enhance national competitiveness and protect national security."

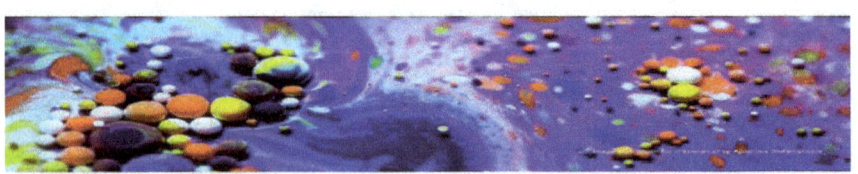

"Military applications of molecular manufacturing have even greater potential than nuclear weapons to radically change the balance of power."

Admiral (Ret) David E. Jeremiah,
former Vice Chairman of Joint Chiefs of Staff *

* "Nanotechnology and Global Security," (Palo Alto, CA; Fourth Foresight Conference on Molecular Nanotechnology), November 1995

Appendix

NGO to Influence IOs

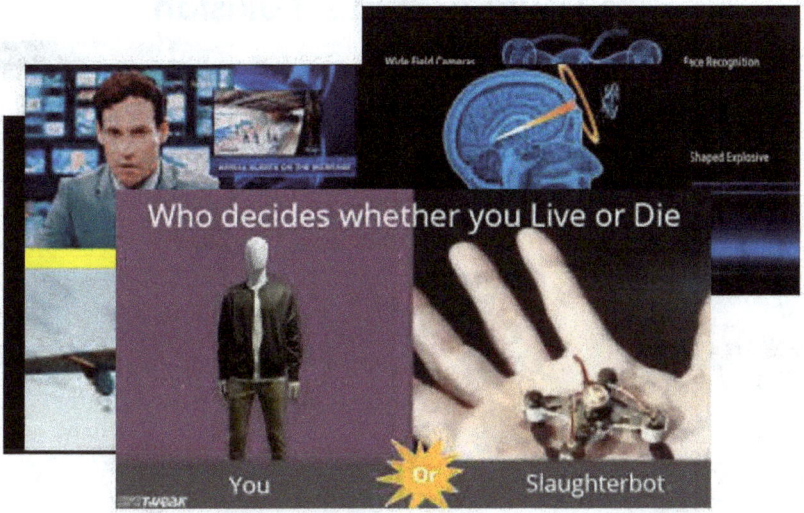

Slaughterbots video: https://www.youtube.com/watch?v=HipTO_7mUOw
https://spectrum.ieee.org/automaton/robotics/military-robots/why-you-shouldnt-fear-slaughterbots

Hype or Prescient Warning?

"This short film is more than just speculation. It shows the results of integrating and miniaturizing technologies that we already have… [AI]'s potential to benefit humanity is enormous, even in defense. But allowing machines to choose to kill humans will be devastating to our security and freedom. Thousands of my fellow researchers agree. We have an opportunity to prevent the future you just saw, but the window to act is closing fast."

– Prof. Stuart Russell, Computer Science, UC-Berkeley

Slaughterbots video: https://www.youtube.com/watch?v=HipTO_7mUOw
https://spectrum.ieee.org/automaton/robotics/military-robots/why-you-shouldnt-fear-slaughterbots

United States Higher Education and National Security

PRC Selling Autonomous Weaponized Drones to Saudi Arabia & Pakistan

https://www.albawaba.com/news/china-selling-autonomous-weaponized-drones-saudi-arabia-and-pakistan-1321951

"Potential Transformative Scenarios"

"1. Supercharged surveillance brings about the end of guerilla warfare."

Greg Allen & Taniel Chan, *Artificial Intelligence and National Security,* Harvard Belfer Center, July 2017, p 31

Creative Countermeasures

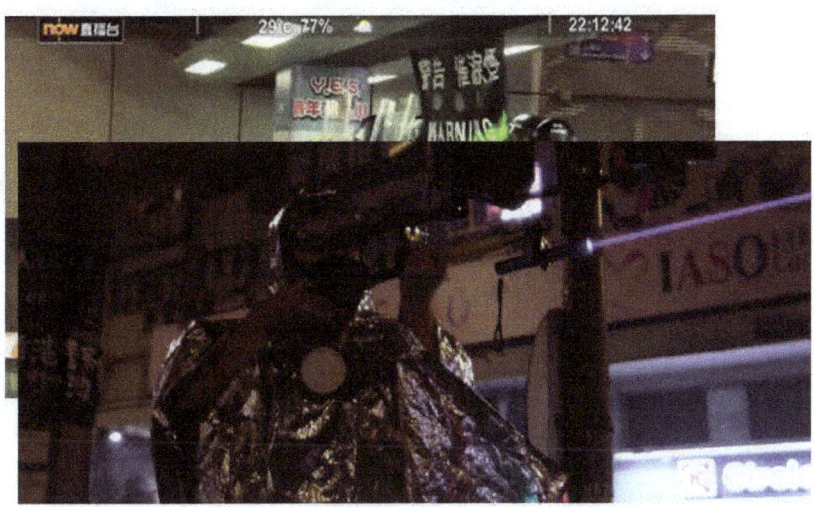

Alessandra Bocchi (@alessabocchi) https://twitter.com/alessabocchi/status/1156513770254012416 29 August 2019, https://twitter.com/hackermaderas/status/1196524155115491328, 18 November 2019

Neurotech

- Global picture
- BCI Types
- First Adopters

Why Study CogSci & Neurotech?

- To understand
  - opportunities for strategic and tactical advantage
    - US & allies
    - Adversaries
  - emerging threats
    - Enhancement & Augmentation
    - Degradation
    - Countermeasures
  - role of S&T fields in military & security-related applications
  - obstacles or enablers from basic research of cog sci & neurotech to deployable applications
  - new deterrence approaches
  - implications for IO, command & control, force postures, organization, & sustainment
  - changes to the character of warfare

2022 Annual Threat Assessment of the U.S. Intelligence Community

"ANOMALOUS HEALTH INCIDENTS

"We continue to closely examine Anomalous Health Incidents (AHIs) and ensure appropriate care for those affected. IC agencies assess with varying levels of confidence that most reported health incidents can be explained by medical conditions, or environmental or technical factors and that it is unlikely that a foreign actor—including Russia—is conducting a sustained, worldwide campaign involving hundreds of incidents without detection. This finding does not change the fact that U.S. personnel are reporting real experiences, nor does it explain every report. The IC continues to actively investigate the AHI issue, focusing particularly on a subset of priority cases for which it has not ruled out any cause, including the possibility that one or more foreign actors were involved."

https://www.dni.gov/index.php/newsroom/reports-publications/reports-publications-2022/item/2279-2022-annual-threat-assessment-of-the-u-s-intelligence-community, 8 March 2022, p. 20

Appendix

Emphasis on Ethics, Legal, & Political Implications to 2050

55	MEDICINE
56	HUMAN PERFORMANCE ENHANCEMENT
62	OFF-WORLD MEDICINE AND BIOTECHNOLOGY
67	BRAIN MACHINE INTERFACES
75	RECOMMENDATIONS AND CONCLUSIONS

Major Global Investments

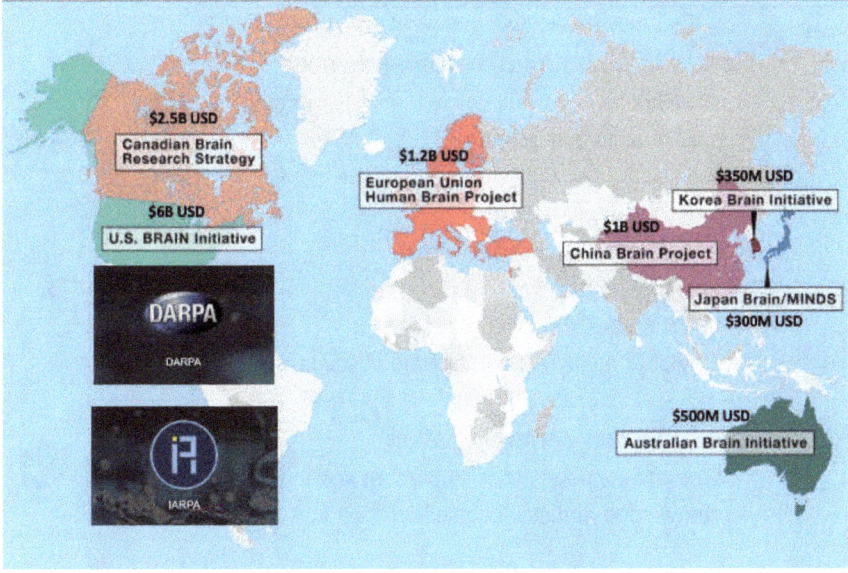

Invasiveness vs Functionality

BCI Applications

- Hands-free control of external devices
  - Prosthetics
  - Cell phones and gaming
  - UAVs

Brain-Computer-Device

- Restoration or enhancement
  - Sensory abilities
  - Monitor attention, stress, etc.
  - Decipher multiple complex data streams

Brain-Computer

- Brain-to-brain communication
  - Remote communication in combat scenarios
  - Knowledge and skill acquisition

Brain-Computer-Brain

Appendix

Will the US or PRC be the 1st Widespread Adopter of BCI Tech?

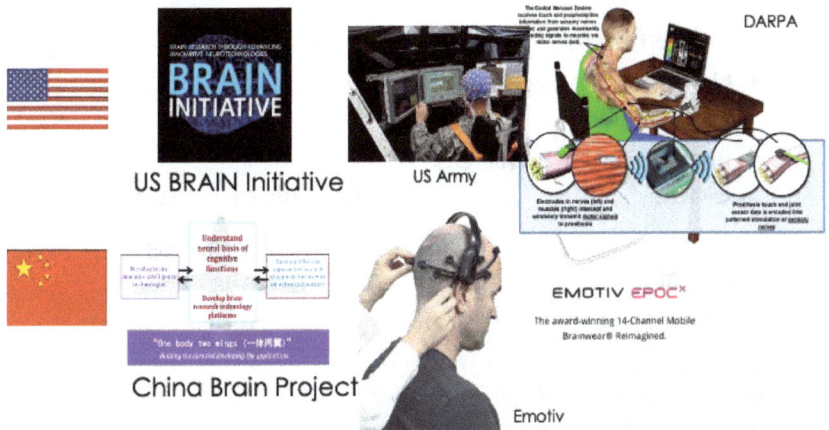

Kosal, M.E., and J. Putney, "Neurotechnology and International Security: Predicting Commercial and Military Adoption of Brain-Computer Interface (BCI) in the US and China," *Politics and the Life Sciences*, 2022, https://www.doi.org/10.1017/pls.2022.2

BCI Adoption Predictors

Qualitative
- Government Structure
- National Innovation System (NIS)
- Brain Project & Military Goals

Semi-Quantitative
- Sociocultural Norms

Quantitative
- Brain Project Funding
- Number of Patents
- Number of Research Monkeys
- BCI Market Share

BCI Adoption Predictors

Political/Institutional

- Government Structure
- National Innovation System (NIS)
- Brain Project & Military Goals

Cultural

- Sociocultural Norms

Economic

- Brain Project Funding
- BCI Market Share

Technical-Material

- Number of Research Monkeys
- Number of Patents

Summary

Nation	Qualitative Variables		Quantitative Variables		
	National Innovation System	Brain Project & Military Goals	Brain Project Funding	# of Research Monkeys	Sociocultural Norms
🇺🇸	• More **robust** innovation system • Focus on **basic research** • History of successful **technology development**	• Focus on basic research and clinical neurotech applications • **Indirectly aligned** with military goals	$6B USD	• 30-40k in breeding colonies • Roughly 60-80% of monkeys used in research imported from China each year	• High individualism scores • Low long-term orientation scores
🇨🇳	• **Developing** innovation system • Focus on **producing already developed tech**	• Focus on clinical applications and **brain-machine intelligence** • **Directly aligned** with military goals	$1B USD	• 300k in breeding colonies	• Low individualism scores • High long-term orientation scores

Appendix

Metamaterials

- What are they?
    - Nanotechnology

- Threat Assessment and implications for the US

Adaptive Camouflage

- Advanced anti – detection technology
    - Biomimetic in design
    - Based on advanced materials & nanotechnology
- Adaptive camouflage '1.0' is already available:
    - IR cloaking
    - Rudimentary clothing projection
    - ... eventually "invisible" tanks
- Exacerbation of the Security Dilemma
- Historically, covert capabilities were major sources of Cold War tension
    - Affect on uncertainty destabilizing
    - Strong potential to enable/encourage covert aggression
- Diffusion
    - Initially, technology will only be accessible to advanced military powers
    - Diffusion to non-state actors pose threats
- Border/illicit trafficking situations
- Detection - counter detection arms race?

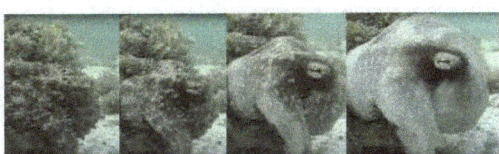

Metamaterials Assessing the Threat

Kosal, M.E. and J.W. Stayton, "Meta-materials: Threat to the Global Status Quo?" in *Disruptive and Game Changing Technologies in Modern Warfare: Development, Use, and Proliferation*, **2019**, pp 135-154, https://www.springer.com/us/book/9783030283414

Summary

- Survey of military applications of emerging tech in the context of how security implications intersect with geopolitics
- Importance of emerging tech globally, especially in the context balance of power, is largely uncontested politically, and simultaneously, it is also an area in need of more robust investigation
    - Seen in rhetoric of political leaders, in civil society efforts advocating for international governance, and in scholarly-level discourse
    - Hype - Balance hope, horror, ... & marketing
    - Social factors may be more important than just funding $$$
- Dual-use
- Much attention given to bias in broader ethics debates but often not considered in security context where human cognitive biases may be different from racial, gender, or other normative biases

Conclusions

- More robust understanding of the role of emerging technologies is needed
  - Technology can exacerbate existing social & political divides (instability)
  - The difference between beneficial and dangerous research is often only one of intent
- Countermeasures understudied ... if noticed at all
- Governance is huge challenge
  - Uncertain tech capabilities
  - International regime challenges
  - "Good neighbor" risks
- Poorly designed and reactionary limitations on research are likely to be ineffective and undermine US and allies' security interests
- Foster pro-active international scientific cooperation as means to encourage beneficial use of technology
  - Pro-active Emerging S&T Cooperative Threat Reduction (CTR)
  - Track 2 diplomacy
- **In the end, it's about people**

Acknowledgements

- Joy Putney, Quantitative Biosciences
- George Tan, Chemistry
- Sathya Balachander, Biology
- Jonathan Huang, Nunn School/INTA
- J. Wes Stayton, Nunn School/INTA

United States Higher Education and National Security

Contact information:

Dr. Margaret E. Kosal
Associate Professor
Sam Nunn School of International Affairs
Georgia Institute of Technology
Atlanta GA

margaret.kosal@inta.gatech.edu
@mekosal

Will AI & ML Replicate Human Biases, Stereotypes, & Prejudices?

Appendix

Racism in Robots

"We project our prejudices onto robots. We treat robots as if they are social actors, as if they have a race."

– Dr. Christoph Bartneck, College of Engineering, University of Canterbury (NZ)

https://www.asme.org/topics-resources/content/racism-runs-deep-even-against-robots

Even here?

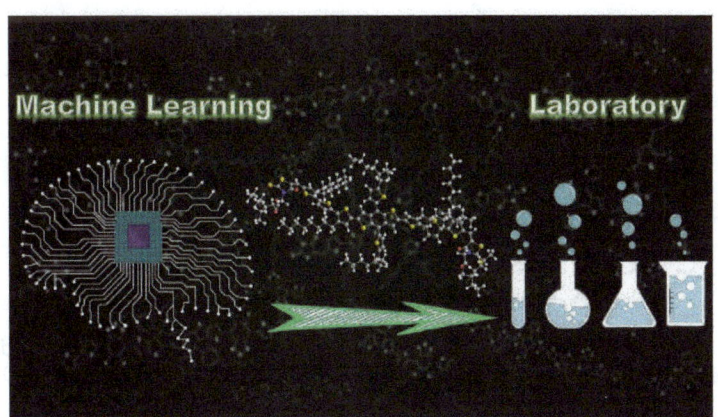

ML-Assisted Molecular Design for Organic Photovoltaic Materials

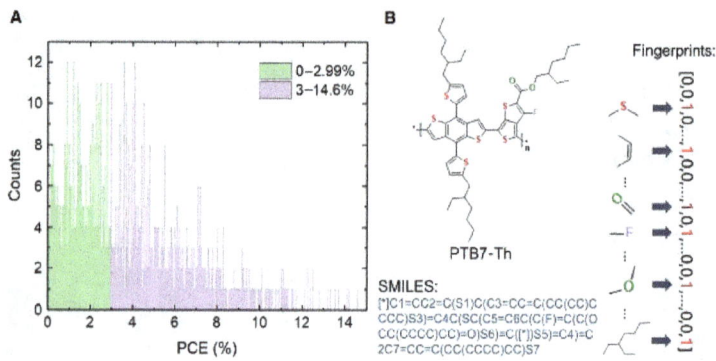

Information about the database of OPV donor materials. (A) Distribution of PCE values of the 1719 molecules in the database. (B) Schematics of expressions of a molecule, including image, simplified molecular-input line-entry system (SMILES), and fingerprints. Credit: Science Advances, doi: 10.1126/sciadv.aay4275
https://phys.org/news/2019-11-machine-learning-assisted-molecular-high-performance-photovoltaic.html

Bias in Chemical Synthesis

"Anthropogenic biases in both the reagent choices and reaction conditions of chemical reaction datasets using a combination of data mining and experiments."

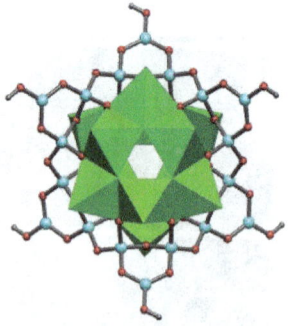

Crystal structure of a vanadium borate complex. Red=O, blue=B, grey=H, green=VO_5

https://cen.acs.org/physical-chemistry/computational-chemistry/Machine-learning-human-bias/97/i36; https://www.nature.com/articles/s41586-019-1540-5

ML for C-WMD

- Limited data sets
- Imagery Analysis
- Power Distribution Analysis
- More useful for nuclear than chemical or bio
- Breakthroughs in AI could allow qualitatively different and better analysis,
  - Determining enemy decision criteria
  - Designing new arms reduction regimes with multiple state equities accounted

ML for C-WMD

	Fully Capable/Feasible
	Mostly Capable/Feasible
	Partially Capable/Feasible
	Not Capable/Feasible
	Not Applicable

	Current			5-10 Years			10-20 Years		
	Data Set	Uncertainty	Value	Data Set	Uncertainty	Value	Data Set	Uncertainty	Value
Raw material acquisition									
Excess production of low-enrichment U-235									
Excess enrichment levels of U-235									
PU-238 short fuel cycle enrichment									
Power generation excess									
Excavations / hidden facilities									
Refusal of international inspections									
Nuclear weapons scientists									
Nuclear weapon design acquisition									
Nuclear Suppliers Group controlled list acquisitions									
Chemical explosives expertise									
Complex electronic expertise									
Delivery system development									
Nuclear testing									
State decision to proliferate									
State decision to employ									

ML for C-WMD

	Current			5-10 Years			10-20 Years		
	Data Set	Uncertainty	Value	Data Set	Uncertainty	Value	Data Set	Uncertainty	Value
Power generation excess									
Excavations / hidden facilities									
Nuclear weapons scientists									
Nuclear Suppliers Group controlled list acquisitions									
State decision to proliferate									
State decision to employ									

Fully Capable/Feasible
Mostly Capable/Feasible
Partially Capable/Feasible
Not Capable/Feasible
Not Applicable

Major Ideational Driver - Uncertainty

- Recognition of the potential applications of a technology and a sense of purpose in exploiting it are far more important than simply having access to it
- Advanced technology is no longer the domain of the few
- Threat of disruptive technologies are of constant concern
- Acknowledgement of amount of uncertainty in assessing future

> "The rapid diffusion of technology, the growth of a multitude of transnational factors, and the consequences of increasing globalization and economic interdependence, have coalesced to create national security challenges remarkable for their complexity." General Charles Krulak, USMC, Commandant (1999)

> "Surprise is what keeps me up at night." General Robert Kehler, US Air Force, Commander USSTRATCOM (2013)

> "What keeps me awake at night is, are we going to miss the next big technological advance? And perhaps an enemy will have that." General Robert Cone, US Army, Commander US Army TRADOC (2014)

Appendix

Changing Strategic Environment

- Post-Cold War *&* post-9-11 international security environment
- Technology & distance no longer guarantees security
  - Changing relationship between science and security
- Globalization and information revolution as drivers
  - Enable spread and accessibility
- Changing characteristics of warfare
  - Peer & near-pear states may still pose existential threats
  - Increasing
  - Asymmetric warfare
  - Interest in unconventional weapons
  - Increased complexity
- Dual-use conundrum
- Multiple emerging (or emerged) disruptive technologies

What once were 2-body problems
have become n-body problems

Weapons Proliferation Pathways
For Identification, Interdiction, & Warning

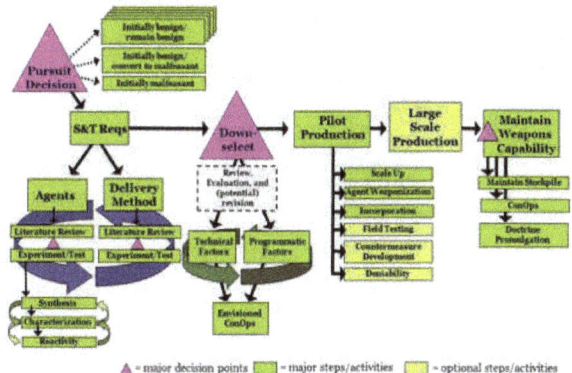

Machine Learning & AI for C-WMD

Student: Pete Exline

Budgets, Organizations, & Cultures

- $4.9B (USD) in FY2021 budget for AI (& machine learning)
- $3.5B DoD
  - $881M DARPA
  - $775M USD R&E (ARO, AFOSR, ONR)
  - $334M DTRA
  - $256 DISA ... (includes $209M for JAIC)
  - $91M SOCOM
- Caveat - for a whole bunch of reasons estimates like this are never perfect but are a reasonable order of magnitude

Appendix

SCIENCE, TECHNOLOGY, AND STRATEGIC ANALYTICS

Dr. C. Anthony Pfaff

Lieutenant Colonel Christopher Lowrance

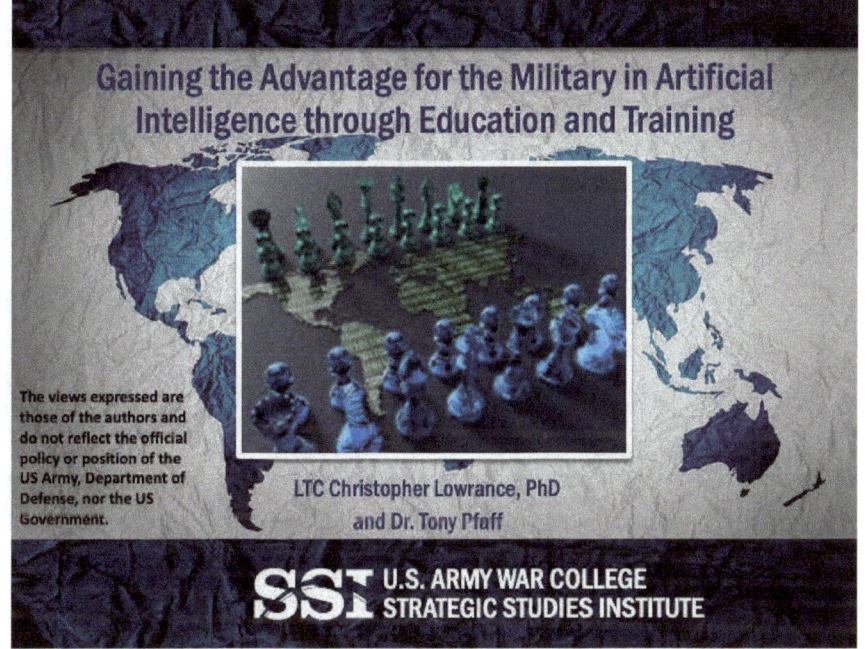

Introduction

"AI will help commanders make sound decisions so much faster that waging war without it will work as well as cavalry charging machine guns on the Western Front."
– Lt. Gen. Groen, Joint Artificial Intelligence Center (JAIC) Director

Research Question:
How can the US gain the competitive advantage in AI for the military?

Thesis:
The US may not have technological superiority in AI, so the US must turn to its *people* for advantage.

Outline

- Introduction
- AI in the Targeting Process
- Challenges
- Overcoming the Challenges
  - Education
  - Training
- Questions

Appendix

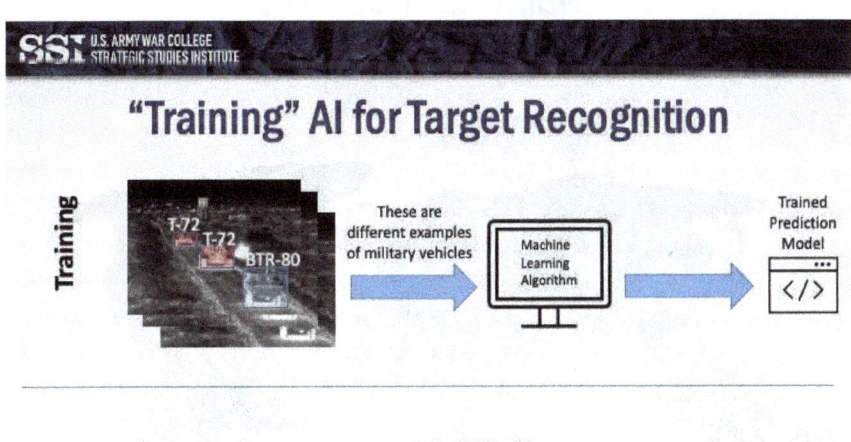

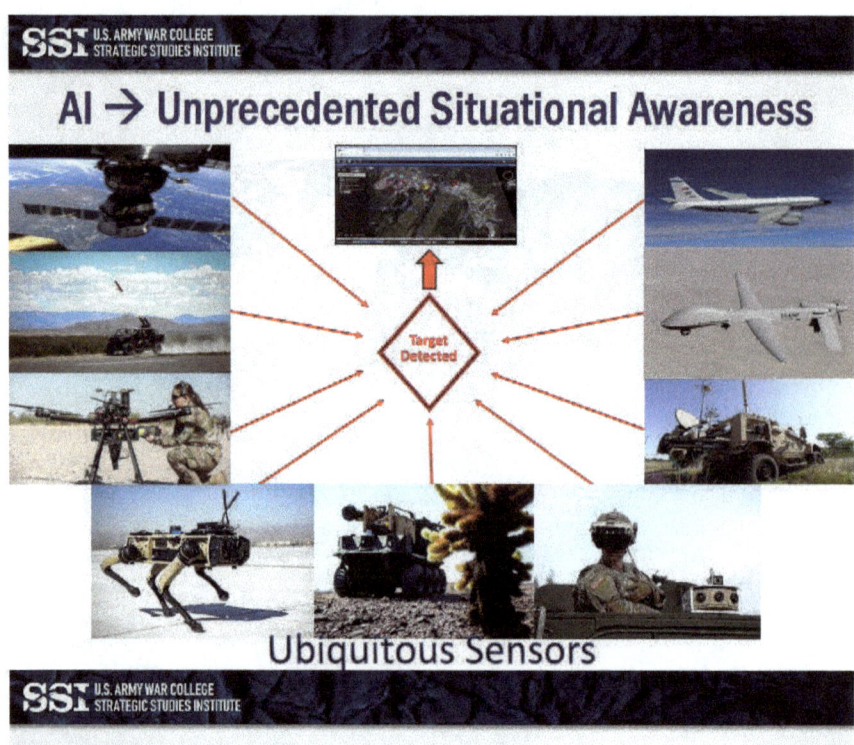
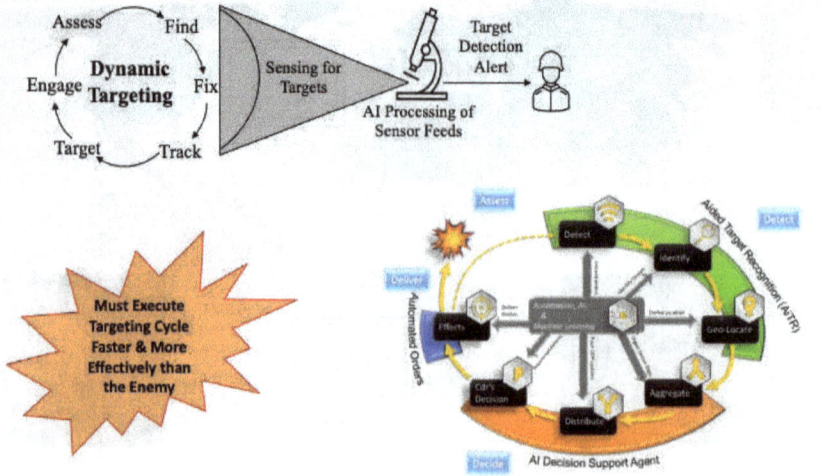

Challenges

Challenges

- The Data Challenge
  - AI requires data to "train" algorithms, but collecting comprehensive datasets that fully represent all possible inputs is nearly intractable
  - Target and environment variability, Lighting conditions, sensor nuances, etc.

- Vulnerability Issues
  - Fooling target recognition systems into not detecting or misclassifying objects
  - Data poisoning

- Performance Issues
  - Misclassifications (AI saying school bus is a missile launcher)
  - False positives (AI saying there is a target, when in fact, there is none)
  - False negatives (AI missing the detection of targets)

The performance of today's AI is a function of the quality and richness of the "training data". When AI encounters inputs that don't closely resemble the training data, then errors can occur, and these errors reduce TRUST as users may perceive the technology as "brittle".

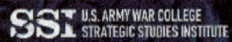

Overcoming the Challenges ->

What will distinguish the US military's AI from its adversaries?

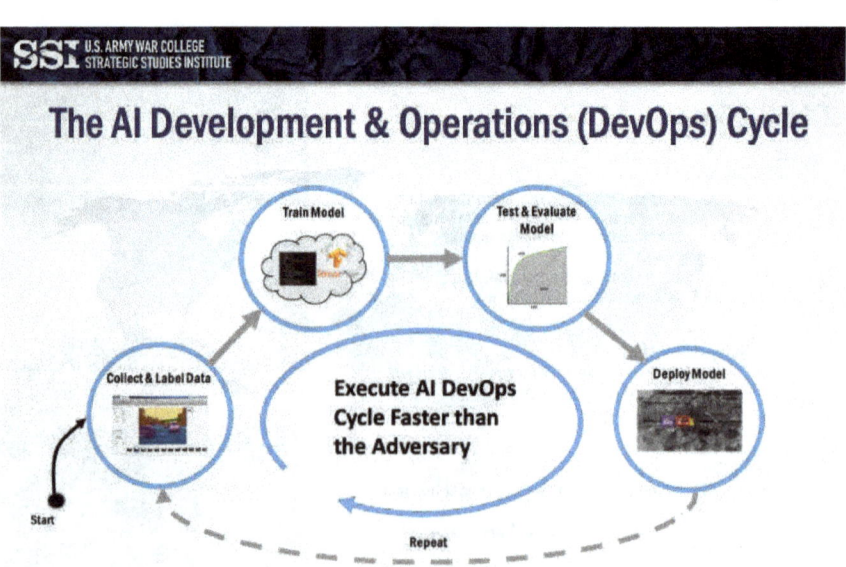

Appendix

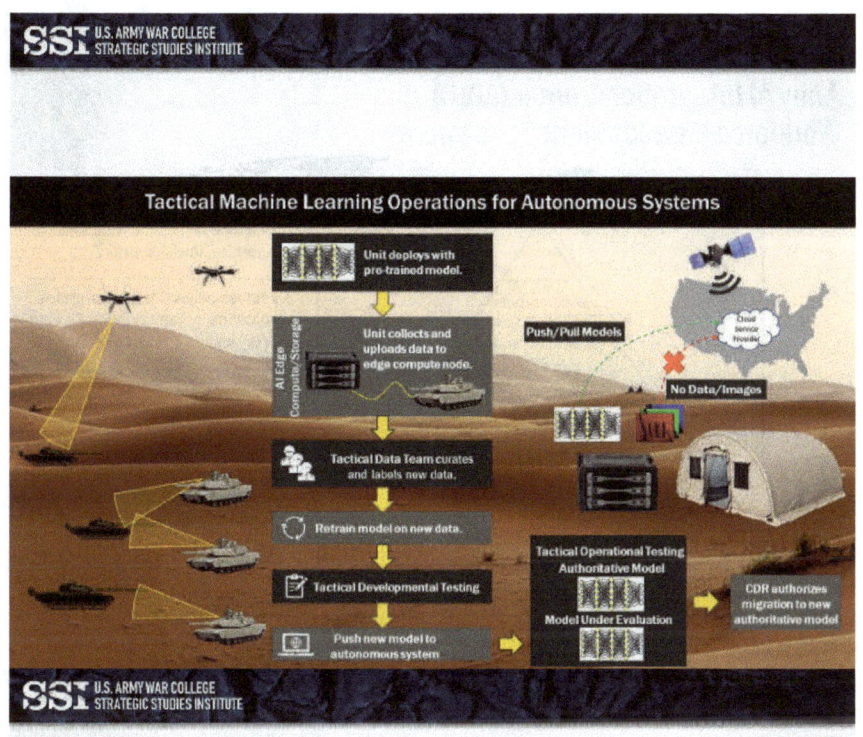

Education

United States Higher Education and National Security

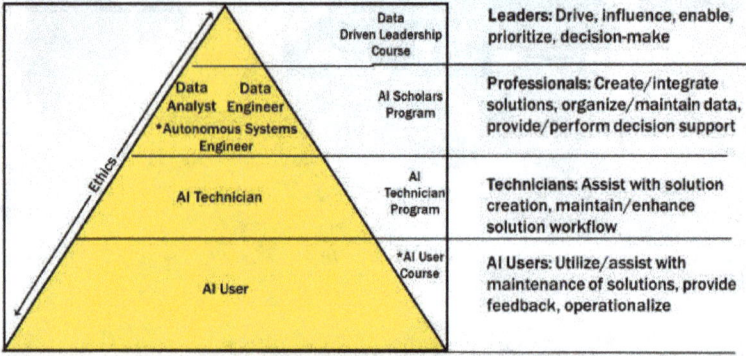

Training

Appendix

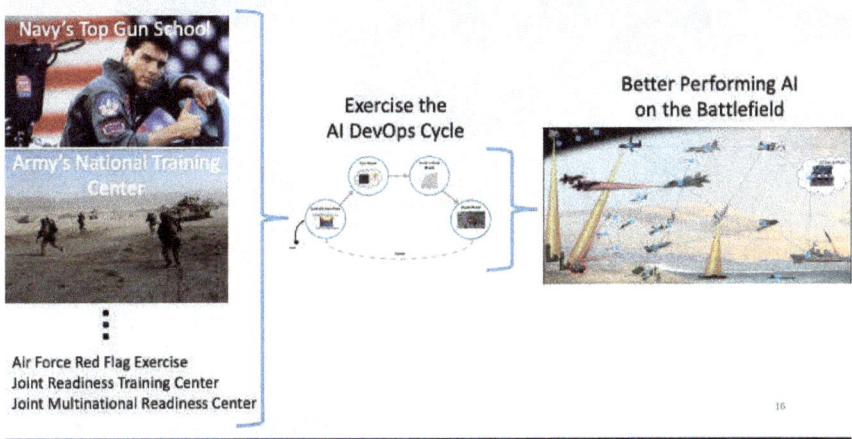

Questions?

The views expressed are those of the authors and do not reflect the official policy or position of the US Army, Department of Defense, nor the US Government.

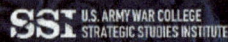

Backup

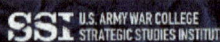

AI2C - Data Driven Leadership Certificate Program

Program Modules

Module 1: Enterprise Data Management (EDM) Foundations
Module 2: Data Strategy and Governance
Module 3: Data Maturity
Module 4: DT and Developing a Data Driven Culture
Module 5: Analytics for Business Intelligence
Module 6: Applied Data Science
Module 7: Optimization
Module 8: Demystifying AI
Module 9: Data Engineering
Module 10: Data Security and Privacy
Module 11: Data Visualization and Communication

The Certificate in Data Driven Leadership covers **key areas – data management, data science, decision making, emerging technology, change management, data privacy and security** – to assist Army leadership in the development of a robust enterprise data management and data science capability to improve decision-making.

Appendix

AI2C Data Analyst – Master's Program

FIRST SEMESTER (FALL)
- Statistics for IT Managers
- Economic Analysis
- Applied Econometrics I
- Object-Oriented Programming in Java
- Database Management
- Digital Transformation
- Decision Making Under Uncertainty

SECOND SEMESTER (SPRING)
- Distributed Systems for Information Systems Management
- Accounting and Finance Foundations
- Machine Learning for Problem Solving
- Supply Chain Management
- Operations Research Implementations
- Data Focused Python
- Applied Econometrics II
- --REQUIRED SUMMER INTERNSHIP--

THIRD SEMESTER (FALL)
- Unstructured Data Analytics
- Advanced Business Analytics
- Decision Analysis and Multi-criteria
- Intro to Artificial Intelligence

FOURTH SEMESTER (SPRING)
- Ethics and Policy of AI
- Data Analytics Capstone Project
- Decision Analysis for Business and Policy
- Exploring and Visualizing Data

AI2C Data Engineer – Master's Program

(5) CORE COURSES COMPLETED DURING FIRST TWO SEMESTERS:
- 11-637 FOUNDATIONS OF COMPUTATIONAL DATA SCIENCE
- 15-619 CLOUD COMPUTING
- 10-601 MACHINE LEARNING
- 05-839 INTERACTIVE DATA SCIENCE
- 11-631 DATA SCIENCE SEMINAR
- SYSTEMS CONCENTRATION

(3) SYSTEMS PROJECT COURSES:
- 15-605 OPERATING SYSTEMS IMPLEMENTATION
- 15-618 PARALLEL COMPUTER ARCHITECTURE & PROGRAMMING
- 15-640 DISTRIBUTED SYSTEMS
- 15-641 COMPUTER NETWORKS
- 15-645 DATABASE SYSTEMS
- 15-712 ADVANCED AND DISTRIBUTED OPERATING SYSTEMS
- 15-719 ADVANCED CLOUD COMPUTING
- 15-721 ADVANCED DATABASES
- 15-746 ADVANCED STORAGE SYSTEMS
- 15-821 MOBILE AND PERVASIVE COMPUTING

Capstone Project

Every student must complete a capstone project that integrates classroom experience with hands-on research. Working alone or as part of a team solving a research problem with either a Carnegie Mellon or industry partner.

Internship

Every student is required to complete an industry internship or adequate practical training. This typically happens in the summer between the first fall and spring semesters.

Electives

Students can take three elective courses. The electives should be any graduate-level, 12-unit course in the School of Computer Science.

AI2C AI Technician Course

Course Overview
- Cloud administration
  - Includes process to provision, orchestrate, scale, manage and monitor cloud services across compute, storage, networking, and security using various cloud interfaces
  - Projects use existing public cloud infrastructure, tools, and services
- Practical Programming with Python
  - Includes types, variables, functions, iteration, conditionals, Python data structures, classes, objects, and modules
  - Learn several IDEs--IDLE, VS Code, Jupyter Notebook, basic I/O operations, and fundamental software development
  - Work on three larger applications - enterprise data manipulation (flat files, data stores), web application backend, and data analysis (visualization, matching)
- Data Engineering
  - Includes ingesting, egressing and transforming data from multiple sources using various technologies, services and tools
  - Develop skills needed to identify and meet data requirements of an organization by designing and implementing systems and data pipelines that manage, monitor and secure the data using the full stack of cloud services
  - Students explore and experiment with various storage abstractions such as SQL and NoSQL databases, data lakes and data warehouses to store, transform and draw insights from data
- Cloud DevOps
  - Design and implement strategies for application and infrastructure that enable continuous integration, continuous testing, continuous delivery, infrastructure as code as well as monitoring.
  - Students will leverage cloud technologies to design and implement solutions to version control, building, testing, release, provisioning, configuration, deployment, and monitoring

SCIENCE, TECHNOLOGY, AND STRATEGIC ANALYTICS

J. Sukarno Mertoguno, Ph.D.

"Synergy between Academia and National Security"

J. Sukarno Mertoguno
Professor,
School of Cybersecurity & Privacy

Appendix

Synergistic relationship of Military and Academia in advancement of technology toward national security

Military:
1) Provides interesting research areas for research to focus on
2) Provides research funding, enabling cutting edge research in the Universities
3) Provides careers in military & DoD civilian (labs, FFRDC, UARC)

Academia:
1) Develops Hi-Tec workforce
2) Develops cutting edge technology through fundamental and applied research

Georgia Tech College of Computing
School of Cybersecurity and Privacy

About Us

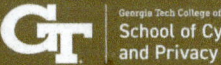
Georgia Tech College of Computing
School of Cybersecurity and Privacy

- Launched Fall 2020
- Today
  - 30 Faculty Members
  - Ph.D.: 60 students
  - MS in Cybersecurity: ~100 students (151 applicants for next year)
  - Online MS in Cybersecurity: ~1200 students
  - Several thousand students are taking our undergraduate classes each year (3,000+ in Introduction to Information Security alone)
  - Undergraduate threads in cybersecurity

- Applicants during start-up year
- Our agility

Relation between higher education and military

- Science and technology is one important factor for high capability and effectiveness of military operation.
- Hi-tech military equipment and infrastructure significantly contribute to the strength of modern military
- Leading US universities are often at the cutting edge of S&T, and often help the military develop future capability, through military research organization such as DARPA, AFOSR, ARO & ONR.
- Military personel with high S&T comprehension can better deals with modern military technology rich equipment and infrastructure
- Exposure to scientific research experience can significantly enhance military personel problem solving capability

The future of that relationship

- Collaboration between academics and military has been cemented with the establishment of:
    - ONR by congress in 1946, and
    - ARPA (DARPA) in 1958.
        The health of these organizations ebbs and floods & is subject to the moods & politics within Pentagon.
    - NSF in 1950 also contributes to the advancement of S&T that leads to national security, purely in the academic realm.
- Strong collaboration between academia and military is mutually beneficial, and strengthen both side for the growth of technology and human capital toward national security

Contemporary issues in U.S. higher education that impact national security

- Relatively low percentage of US citizens pursuing higher degrees (MS & Ph.D.) in S&T, especially at Ph.D. level.

- Programs trying to address the issue:
    - NSF CyberCorps® Scholarship for Service grants (SFS),
    - DoD Cyber Scholarship Program (DoD CySP)
    - DOD SMART scholarship
    - Etc.

What do other countries do to enhance the relationship between their higher education systems and their own national security interests?

- US may be one of the country where the synergy between academia and military (DoD) is relatively more direct and stronger.
- Other such as UK & Australia military also have the synergy, which much smaller budget.
- Many EU research funding are more geared toward collaboration with industries.
- China???

United States Higher Education and National Security

Q&A

SCIENCE, TECHNOLOGY, AND STRATEGIC ANALYTICS
Greg Parlier, Ph.D.

Strategic Analytics for National Defense and International Security

Symposium on National Security and Higher Education
University of North Georgia – Dahlonega
6-7 April 2022
Colonel Greg H. Parlier, USA ret
gparlier@knology.net
256.457.9782

Appendix

Strategic Analytics: Foundational Building Blocks

Analytical Architectures
Information Technologies
Decision Support Systems
The Internet of Things
Engineering Systems
Dynamic Strategic Planning
Engines for Innovation

Capacity, Inventory, and Knowledge

Substitutable Ingredients of System Performance

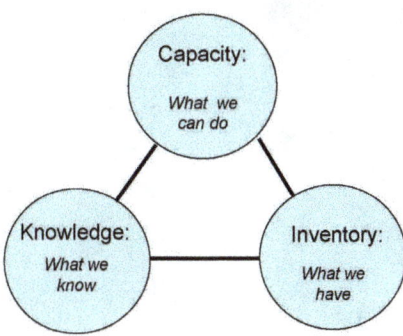

Information Technologies vs. Decision Support Systems

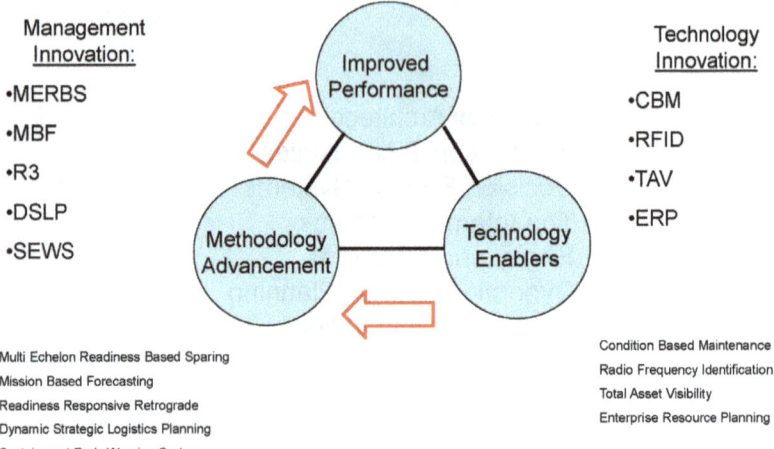

Management Innovation:
- MERBS
- MBF
- R3
- DSLP
- SEWS

Technology Innovation:
- CBM
- RFID
- TAV
- ERP

Multi Echelon Readiness Based Sparing
Mission Based Forecasting
Readiness Responsive Retrograde
Dynamic Strategic Logistics Planning
Sustainment Early Warning System

Condition Based Maintenance
Radio Frequency Identification
Total Asset Visibility
Enterprise Resource Planning

MIT Engineering Systems Division (ESD)

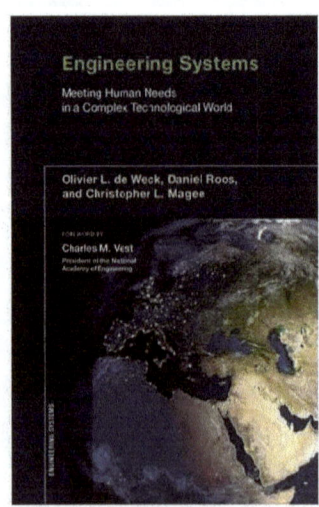

Appendix

Dynamic Strategic Planning

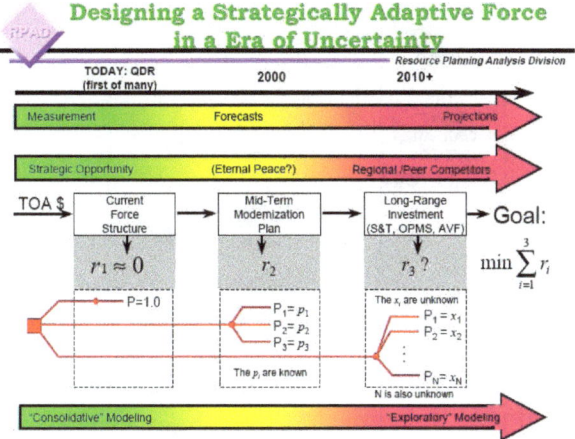

Engines for Innovation

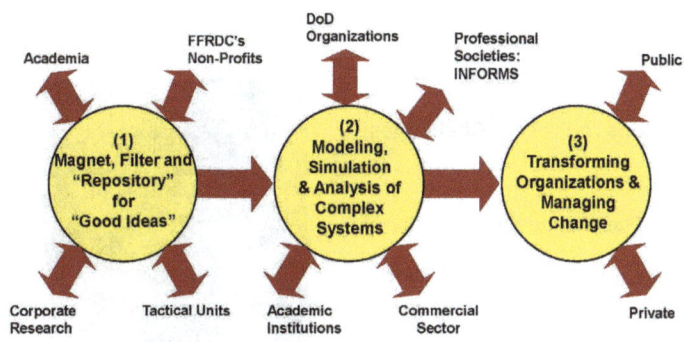

363

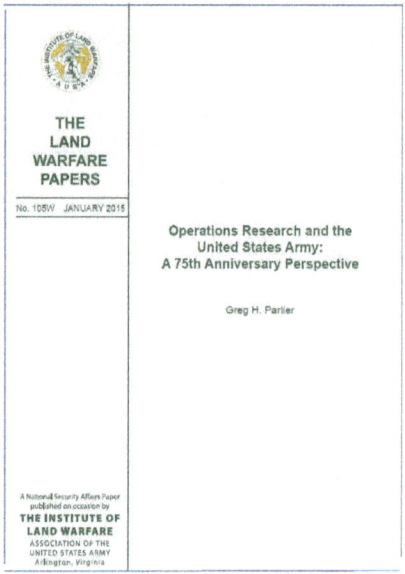

Chapter 20: Strategic Analytics & the Future of Military OR

Applications to Present Challenges:

Resource Planning for National Security

Transforming Defense Supply Chains

Human Capital Enterprise

https://www.taylorfrancis.com/books/e/9780429467219

Public Releasable

Appendix

Reasons for the Book (from Preface):

1. Resurrect traditional Operations Research (OR) for the US Army.
2. Apply "advanced analytics" to our materiel enterprise challenges.
3. Link operational, technical, educational, scientific, and analytical communities.
4. Demonstrate an application of "Strategic Analytics".
5. Document a case study for: analytically-driven, transformational change; a comprehensive, collaborative effort by many contributors.

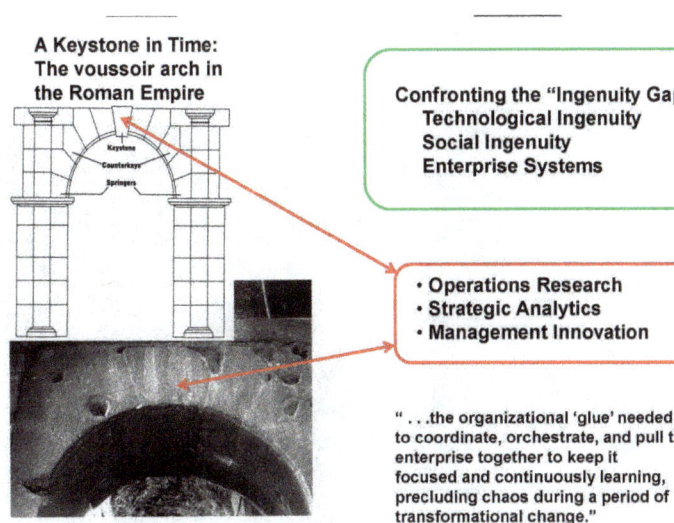

A Keystone in Time: The voussoir arch in the Roman Empire

Confronting the "Ingenuity Gap":
Technological Ingenuity
Social Ingenuity
Enterprise Systems

- Operations Research
- Strategic Analytics
- Management Innovation

"...the organizational 'glue' needed to coordinate, orchestrate, and pull the enterprise together to keep it focused and continuously learning, precluding chaos during a period of transformational change."

Strategic Analytics for National Defense and International Security

Symposium on National Security and Higher Education
University of North Georgia – Dahlonega
6-7 April 2022
Colonel Greg H. Parlier, USA ret
gparlier@knology.net
256.457.9782

LEVERAGING HIGHER EDUCATION TO GROW MILITARY STRATEGISTS

Colonel Francis Park

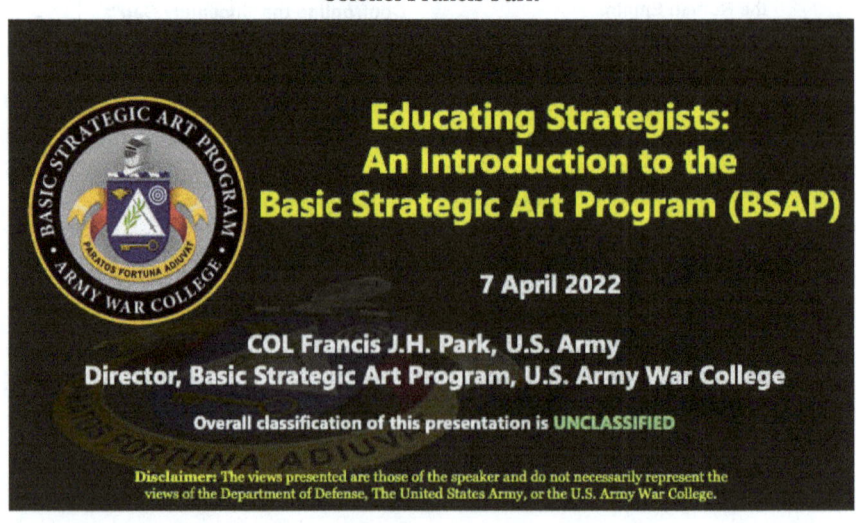

Appendix

BSAP Purpose and Mission
"To inspire and educate strategists to serve the Army and the nation."

Provide officers newly designated into Functional Area 59 (Strategist) an introduction to strategy and to the unique skills, knowledge, and behaviors that provide the foundation for their progressive development as Army strategists.

and . . .

Create a shared common foundational experience

Acculturate officers to the functional area and assist in the creation of their FA59 self-identity

Build the FA59 network

Conceptual Foundations and Standards for FA59

Foundations for development

- Civilian Education
  – Intellectual basis for when doctrine or experience is insufficient

- Professional Military Education
  – Theoretical and professional basis for practical application of strategic art

- Relevant Experience
  – Proficiency to mastery in strategic art

Francis Park, "A Framework for Developing Military Strategists," *Infinity Journal* 5, no. 1 (fall 2015): 9-14

Army FA59 MEL 4 standards

– Master's degree in strategy-related field (12-24 mos), preferably from a civilian institution

– Defense Strategy Course (19 wks DL)
– Intermediate Level Education Common Core (16 wks)
– Basic Strategic Art Program (16 wks)

"One Federal Standard"

Memorandum, HQDA G-3/5, FA59 Accession Policy and MEL Requirements, 6 MAR 2020

Functional Area 59 Developmental Model

DA Pam 600-3 Assignment Categories
- **Institutional Strategy**: plans, strategies, and policy to guide the development of the force
- **Operational Strategy**: plans, strategies, and policy to guide the employment of the force
- **Applied Strategy**: other assignments that require the application of strategic art

Observations
- Skills across planning, resourcing, strategy development, and policy advice are complementary for FA59s
  – Senior FA59s employ all of them
- Development requires a balance across all assignment categories
- A newly inducted FA59 major has about ~10 years/3-4 jobs before facing selection to colonel
- Colonels are slow to learn new skills

BSAP Key Points

- The foundational course to transition basic branch officers to FA59
  – Began operation in 2003, expanded to two classes per year in 2006, three classes per year in 2008
  – Maximum throughput is 48 students per year
- Current instruction up to and including secret collateral (top secret underway)
- Continuous adaptation to account for feedback from the field
  – Informed by continued coordination with FA59 proponent
  – Our graduates must be "ready to fight" upon arrival to their unit (usually a 3/4-star staff)
- Unprepared students can and do face academic attrition
- Faculty are all experienced FA59 officers:
  – Director: division chief (JS J-5), CJTF CIG director, PhD Univ of Kansas, 18 years as FA59
  – Deputy Director: division chief (USASOC), MA George Mason Univ, 14 years as FA59
  – Professor of Strategy: COL(Ret), division chief (JS J-5), PhD Harvard, 13 years as FA59
  – Professor of Planning: COL(Ret), division chief (HQDA), MA Georgia Tech, 16 years as FA59

Appendix

 BSAP Course Overview

Strategic Theory
Theoretical foundations for practice

Strategic Art
Historical case studies of theory/policy

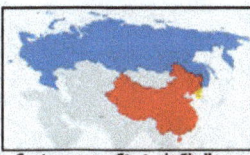
Contemporary Strategic Challenges
Geostrategic context for application

National Security Decision-Making
Case studies on US policy formulation processes

Institutional Strategy and Planning
Design and development of the future force

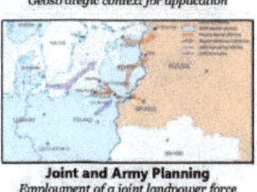
Joint and Army Planning
Employment of a joint landpower force

BSAP prepares Strategists to lead multi-disciplinary groups and facilitate senior leader decision-making by assessing, developing, and articulating policy, strategy, and plans at the national and theater levels.

 Expectations of a BSAP Graduate

BSAP produces FA59 strategists with the basic knowledge and skills that will allow them to critically assess and creatively develop strategic plans and policy:

- Apply intellectual, theoretical, and professional foundations in policy, strategy, and operational art at levels from corps to the national level
- Lead operational planning teams in conceptual and deliberate planning processes
- Produce strategies and plans for theater strategic and national strategic objectives
- Apply knowledge of the institutional Army, Joint Force, and interagency across futures, force generation, campaigning, crisis, contingency, and operations
- Effective communicators, whether written, spoken, or visual

Paratos Fortuna Adiuvat – "Fortune Favors the Prepared"

Backup Slides

 ## Some Propositions on Method in BSAP

Proposition	Implications
• BSAP must provide the most current state of planning efforts and processes in force employment, force development, and force design.	**RELEVANCE AND CURRENCY**
• Experiential learning through exercises and wargames is the most effective way to immerse students in current strategic challenges. • Students must address contemporary issues in an interactive, competitive, peer-led environment.	**PRACTICAL APPLICATION OF THEORY AND DOCTRINE**
• The best way to present current planning efforts is to show, not tell. • Discussing planning experience provides context and insights, but showing the actual products is superior.	**STUDENT LEARNING**
• Planning and strategy development is inherently a social activity. • Seminar discussion and exercises are purposeful training environments for the future operational planning teams and joint planning groups that BSAP graduates will lead.	**PROFESSIONAL DEVELOPMENT**

Appendix

Background: FA59/BSAP evolution

- BSAP pilot program conducted from 2003-2005: One class per year
- From 2005-2015, RA FA59 authorizations expanded from 192 to 441
  - BSAP throughput added to two classes per year in 2006
  - Expansion of program to three classes in 2008
- BSAP stop-gap program conducted at the Institute of World Politics from 2011-2015 as an alternate credentialing source
  - IWP graduates still attended instruction at USAWC to receive instruction on joint campaign planning not taught at IWP
  - Feedback from field was that learning outcomes from IWP were not comparable to those of BSAP and program was abandoned in 2015
- BSAP curriculum expanded from 14 weeks to 16 weeks in 2014 to meet demand for additional planning instruction in the field
- Current annual throughput is 48 (3x16 students), 16 weeks each course

BSAP Course Overview *(16-week course, 3 classes/year, 16 students per class)*
Two military faculty (COL, LTC), two Title 10 civilian faculty, one Title 5 program administrator

Strategic Theory
Foundations of strategic art and operational art
- Evaluations: info paper/brief, two essays

Strategic Art
Historical case studies of military strategy and policy
- Evaluations: info paper/brief, one essay
- Exercise: Civil War 1864-1865

National Security Decision-Making
Case studies on US national security decision-making process
- Evaluations: group brief/senior leader card
- Exercise: NSC Simulation

Contemporary Strategic Challenges
Provide current strategic context to reinforce rest of course
- Evaluations: 15-minute oral board

Institutional Strategy & Planning
Introduce students to the Joint Strategic Planning System and other processes related to force development and force design
- Evaluations: group paper/brief, exam, decision paper
- Exercises: Force Balancing

Joint and Army Planning
Practical application of strategy and operational art through force employment of joint landpower at the theater level
- Evaluations: strategic assessment, planning group work
- Exercises: Theater JFLCC, Landpower, Force Tailoring

Staff Rides
- Gettysburg; introduction to BSAP (one day)
- NCR Staff Ride (4 days); visits to HQDA, JS, OSD, Capitol Hill, interagency (including NSC, State)
- Grant's Overland Campaign (3 days); comprehensive concluding event

Executive Writing: orientation and practica on academic writing and writing for senior leaders

==*Prepares Strategists to lead multi-disciplinary groups and facilitate senior leader decision-making by assessing, developing, and articulating policy, strategy, and plans at the national and theater levels.*==

As of 11 MAR 2022

United States Higher Education and National Security

LEVERAGING HIGHER EDUCATION TO GROW MILITARY STRATEGISTS

Dr. Robert Davis

The Goodpaster Scholars,
or, the U.S. Army's Advanced Strategic Planning and Policy Program

Robert T. Davis, 6 April 2022

The views expressed in this presentation are the personal views of the author and do not represent the official position of the Department of the Army, Army University, of the Command and General Staff College.

Appendix

Bernard Brodie wrote in *War & Politics* that, "As one who helped set up the U.S. National War College by serving on its faculty for the first year of its existence, and who later served on its Board of Advisers—as well as having given lectures there and at the other war college—I feel I can say with confidence that the training afforded at that level is by no means adequate to the needs described. It is undoubtedly a valuable training, and visibly raises the horizons of the officers who pass through it; but as far as changing their basic attitudes is concerned, the training is too brief, too casual, comes too late in life, and keeps the military consorting with each other" [486]

COL(R) Harry P. Ball, PhD, wrote in *Of Responsible Command: A History of the U.S. Army War College*, that "the War College cannot take all credit, not does it need to accept all blame, for graduate performance. A year of War College work is too short to produce judgment, character, and virtue where none existed before, and it comes too late in an officer's life to change dramatically his personality and system of values. At best the bad can be dampened and the good reinforced" [496-497]

United States Higher Education and National Security

GEN Odierno and then
CPT Brendan Gallagher,
West Baghdad, 2007.

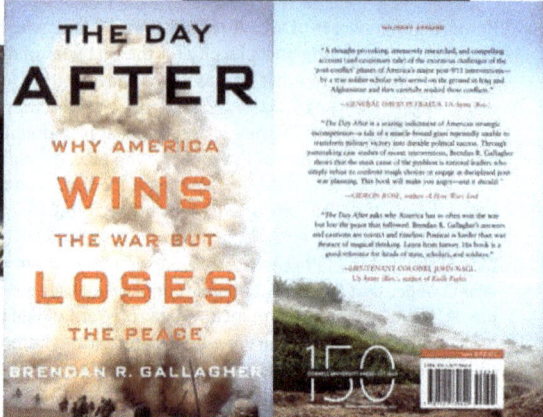

Appendix

EDUCATING STRATEGISTS: AN INTRODUCTION TO THE BASIC STRATEGIC ART PROGRAM

Francis J. H. Park

Basic Strategic Art Program Course Modules

Strategic Theory
Theoretical foundations for practice

Strategic Art
Historical case studies of theory/policy

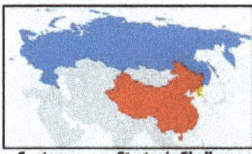

Contemporary Strategic Challenges
Geostrategic context for application

National Security Decision-Making
Case studies on US policy formulation processes

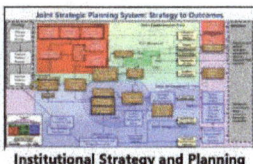

Institutional Strategy and Planning
Design and development of the future force

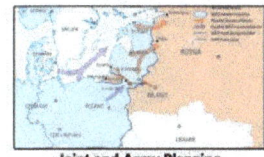

Joint and Army Planning
Employment of a joint landpower force

ENCOURAGING DUAL-ENROLLED STUDENTS TO ENROLL IN CORPS OF CADETS AT SENIOR MILITARY COLLEGES: BARRIERS AND OPPORTUNITIES

Imani Cabell, Ph.D.
Katherine Rose Adams, Ph.D.

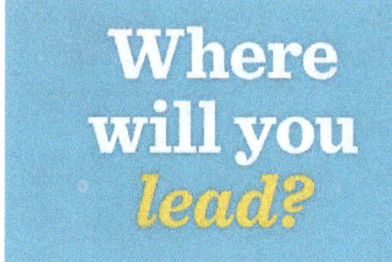 

Encouraging Dual Enrolled Students to Enroll in Corps of Cadets at Senior Military Colleges: Barriers and Opportunities

Presented By: Dr. Katherine Adams & Imani Cabell

Higher Education Landscape

- Falling Enrollment
- Impact of COVID-19
- Generation Z
- HEI Budget Cuts

To navigate the prediction of tighter resources, declined enrollment, and changes in generational perspectives, now is the time to steer recruitment efforts towards new or untapped applicant populations

Appendix

UNG Corps of Cadets
UNG is one of six senior military colleges in the nation.

- UNG & Corps of Cadets founded in 1873, originally chartered as a military college.
- Georgia's first co-educational college and Corps.
- The Corps of Cadets averages 750 (721:FY21) cadets and is consistently ranked as the Nation's best Army ROTC program.
- Corps of Cadets is the opportunity to earn a guaranteed commission in the Armed Forces.
- Produced over 8,000 civilian undergraduate students.

What is Dual Enrollment

Dual Enrollment is a unique program that allows high school juniors or seniors to earn college credit toward a postsecondary certificate, diploma, or degree while simultaneously earning credit toward their high school diploma.

Benefits of Dual Enrollment

 Does not count against 127 HOPE or Zell Miller credit hours cap

 Tuition covered up to 15 hours per semester (for a maximum of 30 hours throughout the program)

 Get a jump start on college

 Immersed in college life, while still having opportunity to enjoy high school activities

 Courses apply to high school and college credit

 Online classes available

History of UNG Dual Enrollment and UNG Corps of Cadets

Dual Enrollment at UNG

- Fall 2013 – 4 Campuses, 253 students (UNG Consolidation)
- Fall 2014 – 4 Campuses, 467 students
- Fall 2015 – 5 Campuses, 627 students
- Fall 2016 – 5 Campuses, 861 students
- Fall 2017 – 5 Campuses, 968 students (Online Dual Enrollment Campus)
- Fall 2018 – 5 Campuses, 1247 students (Online Dual Enrollment Campus)
- Fall 2019 – 5 Campuses, 1370 students (Online Dual Enrollment Campus)
- Fall 2020 – 5 Campuses, 1420 students (Online Dual Enrollment Campus)
- Fall 2021 – 5 Campuses, 1571 students (Online Dual Enrollment Campus)

UNG Dual Enrolled Cadets

UNG first started allowing Dual Enrollment participation Fall 2015

Total Applicants: 7

1- Attended & Graduated UNG

4- Canceled/Incomplete Application

2- Denied

*Some current UNG Cadets have come in with Dual Enrollment Credits from other institutions that have assisted them with the course load and navigation the transition from high school to college.

Appendix

Potential Barriers to Success

- UNG's Corps of Cadets living requirement
- Must be 17 years of age prior to the start of the semester
- Balancing and navigating multiple responsibilities
- Students being able to prioritize their academic work
- Ensuring University System of Ga funding can be applied to military science course
- FROG Week: First week of high school
- Rigor of Cadet freshman schedule
- College readiness and maturity

Recommendations for Success

- Strategically recruit from high schools with greater flexibility or common mission
- Create admission advisor designated to the group for support in completion of application
- Split-Op or Summer Enrollment
- Recruitment by extracurriculars (rifle, JROTC, leadership development)
- Approval of Military Science coursework within a Dual Enrollment track: Hour Adjustments
- The inclusion of HOPE and Miller funding and dual benefits.

United States Higher Education and National Security

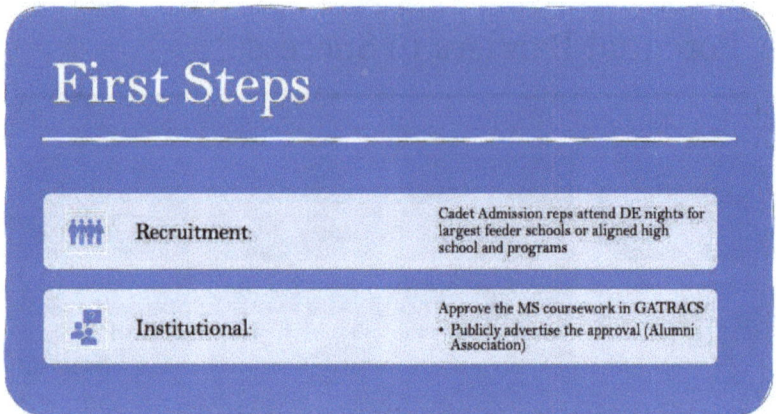

First Steps

Recruitment: Cadet Admission reps attend DE nights for largest feeder schools or aligned high school and programs

Institutional: Approve the MS coursework in GATRACS
- Publicly advertise the approval (Alumni Association)

Courageous Conversations

Thank You

Dr. Katherine Adams
katherine.adams@ung.edu

Imani K. Cabell
Imani.cabell@ung.edu

www.ingramcontent.com/pod-product-compliance
Lightning Source LLC
Chambersburg PA
CBHW050100170426
43198CB00014B/2409